Biography of the Biblical God

Biography of the Biblical God

E. Asamoah-Yaw

Copyright © 2021 by E. Asamoah-Yaw.

ISBN-978-1-6379-0005-5

All rights reserved. No part of this book may be reproduced or transmitted in any form or by any means, electronic or mechanical, including photocopying, recording, or by any information storage and retrieval system, without permission in writing from the copyright owner.

The views expressed in this work are solely those of the author and do not necessarily reflect the views of the publisher, and the publisher hereby disclaims any responsibility for them.

Matchstick Literary
1-888-306-8885
orders@matchliterary.com

Contents

Acknowledgement 9

1. **God** 15
   A General Search for God 15

2. **Why Gods?** 24

3. **The Holy Bible** 30
   i.) Introduction 30
   ii.) Irenaeus (Bishop of Lyons) 32
   iii.) A Brief History 35

4. **Christianity** 43
   i.) Roman Empire and Beyond 43
   ii.) Publication Dates of the
        66 Books of the Holy Bible 51

5. **Who is the Biblical God?** 53
   i.) Book of Genesis (the Cosmic Creator God) 53
   ii.) Book of Exodus (God of Israel) 68
   iii.) Book of Leviticus: 77
   iv.) Book of Numbers: 86
   v.) Book of Deuteronomy (God of Israel) 92
   vi.) Book of Joshua (God of Israel) 96
   vii.) Book of Judges (God of Israel) 101

6. **God According to New Testament**.................................108
   i.) The Book of Matthew (Son of God)....................108
   ii.) The Book of Mark (Jesus Christ, Son of God).....129
   iii.) The Book of *Luke* (the Physician) 133
   iv.) The Book of John (Christ, Word, God) 148

7. **Mary** 155
   i.) Mother of Jesus Or Mother of God?....................155
   ii.) Mary: Mother of Jesus is pregnant.......................157

8. **Jesus of Nazareth**...159
   Who or What was He?..159

9. **Christology and Mariology**..................................170

10. **Mary Magdalene and Jesus Christ**........................174
    i.) What if they were married?..................................174
    ii.) Was Mary Magdalene *Mrs. Jesus Christ?* 175

11. **Questionable Biblical Notes**.......................................183

12. **The Book of Ecclesiastes**...191

13. **Saul: the Apostle Paul** 199

14. **Conclusion** 203
    Bibliography 211
    Index...213

DEDICATION

This work is dedicated to all those who see God as a non-divine entity: To all those who see God as something within 'the individual self' and not as a spirit or something outside the self. And to those who identify the human brain, a womb-manufactured organ, as the ultimate God, which controls all humans at all times everywhere.

Acknowledgement

There are certainly not many people out there who are willing to listen to any challenge of the existence of spirits, ghosts, witches, angels, demons, or gods. Christians in particular cherish their God to the extent that any criticism of His existence sounds like questioning something that is universally accepted as fact.

About sixty years ago, I used to ask my two grandfathers some puzzling questions which I am still asking in the year 2011. My grandfather on my father's side was Reverend Nana Amos Asamoah, who was the chief priest of the District Methodist Church, and the other was Nana Kwaku Manu, on my mother's side, who was similarly a leading member of the local shrine or religious organization. Such questions as where babies come from and where dead people live had bothered me for more than half a century.

Being their first grandchild, the sight of me always reminded them of my usual childish questions. Both of them actually made songs with my questions in their respective religious tunes which sounded very pleasing to my ears. They always sang the same songs when they saw me running to their arms; but their answers to my questions were never satisfactory. My Christian grandfather told me that children came from God who lives in Heaven. While my Fetish religious grandfather told me children came from Otwiadeammpong, who lived in a big rocky cave located in a thick forest far, far away from the village. Both of them told me that dead people are not really dead; rather, their souls or spirits are still alive in a different world around us. They see us, but we cannot see them. My fetish grandfather added that some of the dead lived in the same thick forest where children come from; that he himself has personally seen some there. My other Christian grandfather, Nana Amos, however, told me not to be afraid of ghosts because the Heavenly God is constantly watching and protecting all of us from evil spirits. I was naturally afraid whenever I was alone; yet I remained very curious to find out where exactly these invisible people lived.

Nana Amos Asamoah had given me one of his treasured properties as a gift, the Holy Bible, way back when I was a young teenager on the day he baptised me, his own grandchild, in his own Methodist Church. You can imagine the joy. A few years later, when both Grandfathers and my father had died, my mom moved us from the village to a big town nearby, only three miles away, and enrolled me in the Seventh Day Adventist Church School. Church attendance there was mandatory. Absence from church was the most serious crime. After another four years, my step father moved me to Presbyterian Church Mission School to complete my Middle School education. Bible reading was a must and inescapable. Unfortunately, for me at the time, my girl friend was a Catholic, and I was compelled to join her occasionally to Catholic Church services as well on most Sundays.

After Middle School graduation in 1958, I stopped attending churches completely, but concentrated on reading the Holy Bible out of curiosity. The year was 2004, when I decided to retire early and devote more time to search for the biblical God of my grandfather Amos, and the local cultural God, the Otwiadeammpong, of my fetish grandfather, Nana Kwaku Manu.

In March 2012, I will be seventy years old, and I am still asking the same questions I used to ask when I was a child with innocent immature mind. My regular friends are somehow used to my usual religious questions though this time the questions are complicated. But perfect strangers usually think I am either drunk or crazy, whenever I talk to them about the non-existence of God of any kind. There are no ghosts, no angels, no demons, and no spirits of any kind in reality in our universe. All of them exist in people's mind. They are nothing but man-made entities which human brains perceive as real. They are mere illusions.

After completing this book's manuscript, only a handful of friends agreed to proof-read my work. One very well-educated Christian friend read the first twelve pages and sadly refused to read further. He called me to say that the content of the book was contrary to his religious belief. He is not willing to abandon his faith; hence he could not read beyond what he had read. He does not want me to mention his name in connection with this book. I will simply thank Mr. Kwart of Ksi for his candidness. Mr. Tom Sawyer too did excellent work for helping to cross my [t]s and dot my [i]s. Most of his suggestions brought useful improvements to the overall structure and ideas into proper perspectives; Nana Amo, thanks a million, but don't stop your routine morning prayers—as you threaten to do after reading the manuscript—if prayers give you comfort.

Eno Ntiamoah Opoku Boamah was the first, the youngest and the only female reader to finish reading the manuscript and the only one who wrote a lengthy critique with several of my mistakes corrected. She is also a Christian, but she assured me that she will read the Holy Bible this time with a clear conscience. She wonders if she should continue to go to church after reading this work.

Eno, personally, it is a waste of precious time, but you will be a better judge of that.

Justice Debrah also read the discourse in two weeks' time, and he warned me not to publish the book, because it questions religious people's faith. His highness nevertheless does not reject my appeal to personal faith as opposed to divine faith. My learned friend, Mr. Bramfi, deserves many sincere thanks for devoting his precious time as an accountant to sit down and argue on salient points of the book with me. His critical mind in religious matters was one thing I needed very much to sharpen my interpretations of illusive biblical concepts.

How can I forget Nana Osei for his incredible knowledge of biblical ideas? Nana's perspectives of the New Testament, his understanding of Greek and Hebrew words, plus his knowledge of Martin Luther's role in Christianity gave me a smooth ride in sailing across the middle ages to the modern theological school of thought. I am particularly greatly honored for the hours and days he devoted to make this book become a reality. Nana had to drive about six miles to my house to sort out differences in literal and theological definitions of biblical words and phrases. Nana, all I can say is simply, thank you.

The same appreciation extends to my numerous casual acquaintances in my locality. Apart from reading and writing, walking is my favorite hobby. And I like to thank all the innocent people who simply responded to my greetings with a friendly religious phrase like 'by Gods grace', not anticipating the saying could become a hot debate between us, about who that God is. They are indeed the ones who inspired me the most to search deeper into the core meaning of godly ideas which can today be seen and read as a book. I have always challenged people to explain what they mean whenever God's name is mentioned.

Above all, I must thank my wife Mrs. Juliana Asamoah-Yaw, who I always introduce as my boss, my land-lady, and my accountant, for her endless support in bringing this work into perfect fruition. She was never tired of asking questions about spirituality and the creator of the Creator and their place of residence if there is such an entity. She was the one who primarily ignited my brain cells, for thirty-five years, to look for answers to the numerous biblical illusions intentionally created by ancient theologians. For the eight years I spent in reading and surfing the Internet to put together these biblical opinions, there was never a day when I missed her support. Many, many thanks to her and to all others I could not mention here. The one amazing observation which is worth mentioning here is the reluctance among people with decorated University laurels to face challenges head-on. They simply prefer not to discuss religious faith. Why?

Lastly, I like to express my sincere appreciation of the wonderful professional work done by the Xlibris publishing company staff who transformed my manuscript into what it is today. Nana Yaw, also known as Alexander Agye-Ampofo, Kumasi was the last person who assisted me with his advance knowledge in computer transmission of the manuscript to the Publishers. Many thanks to dozens others I could not mention their names here.

Chapter 1

God

> Then God commanded. And now *we* will make human beings; they will be like *us* and resemble *us*.
> (Genesis 1: 26)

A General Search for God

Throughout human history and among all ethnic cultures on planet Earth, there has always been a doubt as to whether the word 'god' has a special meaning or not. Many have wondered whether it is just a mere word or one with deeper significance. The word is nearly always written with a capital letter 'G' as though it is a proper noun Again, in some senses, the word becomes a term, a notion, and indeed a philosophy representing a spiritual idea or a set of ideas. Question is, what or who is God and why?

It would seem appropriate to begin our search for a better understanding of the word from a literary source. For example, the Webster's Ninth New Collegiate Dictionary defines God with small 'g' as 'the supreme or the ultimate reality; as a: Being perfect in power, wisdom and goodness that men worship as creator and ruler of the universe.'

In this literal sense, therefore, god is a name given to the originator, the ultimate designer, or the owner of everything in the universe, including both living organisms and non-living things. By this first literal definition, the word should be written with capital 'G' because it represents a name, a being, or an object believed to have more than natural attributes and power, and to require man's worship; a person or a thing with supreme value; a powerful ruler, an object, or a deity. The dictionary simply defines the word as it is popularly understood. It does not agree or disagree of god's existence.

The overall meaning of god, simply, therefore is whatever an individual person or a group of persons perceive as god or as the most revered supreme entity.

God, therefore, is a perception and necessarily a condition of the mind. It is something that exists only in the mind. It is not an entity that can be said to occupy a space somewhere at any given time. The human organ known as the eyes can *look* for a god and may not necessarily *see* him or her or it; or touch him or her, or it. *God can be seen only in the mind by imagination.* By applying the above explanation of god, there are necessarily as many gods as there are people in the universe. It is the mind or the accumulation of acquired information in the individual's brain *organ* that knows who or what a god is. It is the brain that processes the image of God for eyes to see, mouth to speak, and the ear to listen to the nature or characteristics of the perceived God.

For without the brain the eyes merely look, the mouth merely talks, and the ears merely hear.

It is written in religious texts that God is everywhere. Yes, God must be everywhere because wherever you are, your brain is with you—even when you are sealed in a bottle, you cannot leave your brain behind. You can definitely condition your mind that God is with you even in the sealed bottle. To get out of the bottle, you will have to engage the brain to process its stored information to find an escape route. You may pray to God for help, yet the prayers would still be an action of the same brain using the acquired religious information stored in your brain cell. Mentioning God's name may increase your comfort level in the bottle, but you may die in the bottle if the state of your mind at the time is deranged or anesthetised in such a way that you lose focus of your whereabouts. At that time, your link with God will be zeroed out. In this analogy, a religious person and non-religious person will end up the same way if the human brain is disengaged or out of focus. My point here is this: Why flatter yourself with something invisible like God but not directly vigorously engage your brain power for a calculated result? Because at the end of the day, deep down in you, you realise that you are your own God. Your brain, which is your god, and you, are inseparable. The power of your brain or god will depend upon the quantity and quality of information stored in the brain organ. There are, therefore, as many gods as there are humans from individuals' perspective.

Historically, however, human beings have always lived in groups of families, communities, societies, tribes, cultures, states, and countries; each group member tends to possess a common trait among each other. Thus, through blood links and or socialisation process of group members, *human acceptance of the nature of god tends to be alike amongst every cultural group.* The group members develop faith in their god. They create symbols, methods of worship, requisite paraphernalia, and laws. Obedience is enforced by elders or the faith leaders in accordance with the rules and regulations made by the leaders but alleged to have been dictated spiritually by their god. Followers are coerced to pursue the creed orders. Their god becomes part of the overall *de facto* political, judicial, and social machinery through whom all powers are derived.

In most religions, the founder's name becomes the accepted name of the creed. For instance, the legendary biblical Moses was allegedly appointed by the Jewish God to lead the Israelites to the Promised Land. It was through Moses that God delivered rules and regulations or the Ten Commandments, which guided the Jewish descendants of Abraham, Isaac, and Jacob according to the Bible. The New Testament called them the Laws of Moses, not God's laws.

It is estimated by the United Nations Organisation, for example, that there exist over 7000 human cultures in the world. By implication, there exist over seven thousand different gods or established traditional religious institutions throughout the world.

Universally, every culture without exception has a unique name for its god. These traditional names may have been lost or evaporated today as irrelevant. They may not be currently functional, but the names still exist linguistically in all cultures.

Fortunately or not, historically, dominant cultures tend to militarily conquer weaker cultures and subsequently suppress the weaker ones by demeaning their gods or faith systems as inferior, and in its place establish the dominant culture's faith system. In most instances a few generations later, through coercion, tactical persuasion, indoctrination, and human adaptation instincts, the foreign faith becomes the norm. Many generations later, some elements of the forgotten classic faith become apparent in poetry, idioms, literatures, and art.

In the primitive days the world population was very small, and human knowledge of organic matter and the environment was equally small. Human understanding of nature was mostly dependent on superstition, myth, and imprecise primitive science. As global population increases so does human knowledge of nature. Dependence on myths, superstitions, and adaptation of different foreign mythical rules expands beyond ethnic boundaries.

Over ten thousand years ago, for instance, our ancestors worshiped numerous gods for several reasons; for example, there was a god of the sun, god of the moon, god of rain, god of night, god of wind, god of storm, god of justice, god of the earth, god of heavens, god of wealth, animal and plant gods, and sea and river gods; there was even a sneezing god and indeed thousands of other gods. These gods were the imaginations of human brain, perceived and depended on by humans because of our natural survival instincts: fear of death, need for external perpetual protection and pursuit good life.

There is no doubt that in the mind of ancient people, these gods served useful socio-economic and political purposes at the time they were created. They equally created indelible problems for generations thereafter: the problem of constant reflection on life after death, heaven and hell, angles and demons, ghosts and witches—all these ideas were brought forth by our human ancestors. In our daily practical life, none of the above appears to be real, yet we cannot stop thinking about them. We simply weave these ideas into our everyday behavior and perceive them as though they were real.

These ideas become permanent faith and later into spiritual faith, and transform to a theology under the supervision of institutionalised and regulated religious organisations, usually represented by churches, parishes, and other such bodies.

The inherited faith (mostly foreign) that has been passed on to present generation will also be transferred to future generations, yet for fear of the unknown, as stated above, most people reluctantly remain loyal and blind faith followers regardless of the faith being ancestral or foreign.

It is estimated by the UN (2010 Year Book) that more than 60 per cent of world population today believe and worship ancient gods. And there are about two billion followers of Christ out of the seven billion world population as of September 2011.

Although modern science and technology have abundant verifiable evidence that challenge God's authenticity, the existence of god in the minds of people remains deep-seated.

It is very difficult, if not impossible, to tell a religious person that Gods are delusions, that they are so because the idea exists only in the human mind, and that they have no real existence. The practices have always been dream—and magic-based, and no realistic substance. Subconsciously or not, religious leaders with such titles as prophets, messiahs, deacons, priests, reverends, fathers, rabbis, imams, maharishis, gurus, and several frightening identities often lead people with hoax promises in Gods' name. In reality, this has always been an effective tool to freeze the mind by instilling in the human brain a belief system which merely *appears* to guarantee success and happiness.

Reliance on divine faiths or spirits always leads people to follow fraudulent artistic leaders who claim to be messengers of God: leaders whose real religious intention is to make money. Such names as Jesus Christ and God of Israel have become a major economic tool to make money without sweat in poor neighborhoods around the world. Success is possible only through personal efforts. Absolute divine faith corrupts and freezes the mind absolutely. Faith in gods and spirits prevent the mind to critically examine the individual self. Faith in god diminishes followers' efforts in investing faith in their natural capabilities. Faith in God and spirits guarantees ignorance of the self and assure transfer of power, wealth, and self-esteem from followers to leaders.

It is self-evident that belief in the self, the personal qualities or capabilities of any individual, can lead any ambitious person to accomplish any conceivable goal without god's interference. Millions of people for millions of years have hopelessly relied on divine magic, with a hope that miracle is possible without personal efforts. It is simply fallacious. Faith from within, not outside, is the key to success. Just convince yourself that you can, then make a serious move with perseverance, commit yourself fully, and you can overcome all difficulties.

There is nothing, absolutely nothing, impossible in this world for the human brain. It may take sometime to conceptualise and realise any dream; yes, but it is only a question of time and serious commitment, not its plausibility. It is the brain and nothing but the human brain that can make or unmake anything, including making and unmaking god and human beings as explained above. The question is this: Do humans of today's scientific world need a God if we are *inwardly focused*? No. But if your answer is yes, please ponder for a moment and ask, 'For what?'

A religious fanatic may ask, 'Is it not God who made the human brain?' Again the answer is no; it has nothing to do with God or any divine influence. While modern science is vigorously trying to unlock human potential, religion is struggling very hard to keep our brains locked up in the vault of divine faith.

Our maker is our mother. Your mother could have willfully, voluntarily, and individually, without consulting your father or anyone, decided to terminate the pregnancy at any time during the entire nine-month period. And you would have had no existence when the fetus is aborted. This is the natural gospel truth which religious people refuse to accept. Your creator or maker must necessarily be your mother.

Your God or your supreme creator is necessarily your brain.

Your heart is the first living organ that begins your creation or metamorphosis in the uterus. The brain forms during the first three-month period in the womb. And it is *only* the female carrier of the pregnancy who can stop the genesis (organogenesis) or development of the brain organ inside the fetus to its full maturity. This is a scientific fact. Human brain is not an artifact, yet its creation and formation is through human activity inside the pregnant female body and therefore a human made, not God made.

There can be an infinite number of questions about the formation of human beings, most of which, if not all, are currently scientifically explicable. To all those who are anxiously looking for the maker of the first human, please re-read your man-made scriptures carefully without prejudice, and I guarantee that you will look elsewhere for your maker. Religious fanatics' explanations of our maker are awfully untenable in our modern scientific world.

They are obsolete hypothesis. The first human couldn't be a divine creation because of its complexity, if *in fact* there was *ever* such a thing as the first human being. The *first*, as a concept in humans, is a biblical idea. There could never have been the first man, the first woman, first monkey, the first bird, the first fish, the first tree, or the first in any natural thing.

The globe is too massive for it to be preoccupied with just one unit of everything to occupy it, and then later procreate into large numbers. The variation in climate and zonal living conditions, the soil texture, and the thousand other variables make the biblical 'first' concept an illusion. Natural laws of our universe or of our cosmic system have made possible the contents of planet Earth. It is logically impossible to use the word creator to explain anything that is not man-made. Modern science in physics, biochemistry, and space has sufficient data to explain the mechanics of this universe.

Admittedly, human beings are conservative by nature, and always reluctant to change, especially so when beyond their fortieth year. Yet it is the same period in life when we begin to either question the status quo doctrines or simply reluctantly totally grab the doctrines in preparation for our supposed lives after death—plus a perception that pursuit of righteousness may pave way for us to Heaven or to somewhere for everlasting life.

Again what is Heaven or Hell? Is it real or imaginary? Think about it! And if your brain tells you that it is real, ask yourself for evidence of its reality. Evidently, human recorded history for millions of years shows that *none has ever returned after death to testify its reality*. If, however, you answer yourself no, then you may conclude with the final question, 'Why is it still in my mind if there is no such thing or place called Heaven or Hell'?

Well, because of life's uncertainties, and the brain organ is the main perceptive and protective engine of the entire body, the brain secures all information (including religious doctrines) which is *presumed* to shield the body in the memory cells permanently. If your brain processes a religious faith as the most effective safety device to protect the human body, you will by all means opt for that faith. It does not necessarily mean that your choice is good. No, it merely means that based on the amount of religious information stored in the brain, the best answer at that moment is what your brain will process for you.

The brain can only give you what it has been exposed to. Your brain's exposure to superstitious information will bring you only superstitious perceptions. If you later expose the brain to scientific information, your brain will necessarily process both information and give its best answer based on the quality of both information it is exposed to. If our family members and friends are Christians and the schools and universities we attend is of the same faith, our brains will primarily resonate towards Christianity on questions of faith. And having been born and bred with Christian information stored in the brain cells, we automatically address ourselves as Christians. But the question is, 'Are we? Do we really understand the basic meaning of God or Christ?'

We are about to get on board a vehicle to begin a conducted tour of the Holy Bible city in search of God, who we are told is our maker. Fasten your seat belt now!

Chapter 2

Why Gods?

Generally speaking, human beings do not voluntarily choose to perish or choose to commit suicide, although the choice to die belongs to every individual. Truthfully speaking, every person can set up a date and time when she or he wants to die; yet, we all choose not to commit suicide but rather to procreate and also struggle to live as long as possible. We all wait for something to kill us.

Empirically though, all organisms have a natural desire and instinct or impulse to perpetuate the self through personal care, family dependency, and the general public defense mechanisms. Struggle for continued existence by all living organisms necessarily require some kind of life assurance or protection against the unforeseen or unpredictable events that threaten life. In the past, this inherent organic desire in humans was seen as the will of the perceived god. Thinkers and theologians of the past in some cultures had written down these opinions as sacred godly truth, without supporting verifiable evidence. They disregard human efforts.

It is always claimed by believers that trust or faith in god is the answer to everything. How about reliance or faith in yourself instead? The self or the brain power needs to be exploited *fully*.

We are our brains. The brain becomes our maker and our keeper as soon as human nurturing begins.

All cultures create myths in an attempt to explain causes and effects of every inexplicable human experience as the work of the prevailing cultural God or Creator. We, however, do not search for proof, because through nurture, God or the Creator is responsible for all cosmic mysteries. The absence of indisputable evidence in support of these ancient claims in effect does not astonish anyone because that is the norm in the culture. Many people are still clinched to ancient religious faiths today because it has been inherited from their ancestors and they see no reason to question it.

Many believe that all human beings must have faith. Yes, this is true, but the faith in question *should not, and I repeat should not* necessarily be a *divine faith or spiritual faith*. We are all 'divine faith addicts.' We see no reason to question our indulgence in this practice because it appears to offer us some survival ingredients in life, such as food, water, hope, comfort, pleasure, security, and happiness. But at a close examination, these show up as mere appearances. The divine faith will not and cannot make these available if, for instance, we keep praying non-stop for millions of years. Prayers don't count. What counts most is your brain. The interesting aspect of these purposes of life is that they are all based on our mindset. One needs not depend on a supernatural being to have a meaningful sense of life. Rather, the acquired knowledge which is stored in the brain cells is the entity which must be processed and utilised to accomplish everything. Failure to accomplish a goal means that you do not posses enough or quality information necessary for your brain to process. The power of the almighty brain organ is the thing that matters, not an invisible non-verifiable entity such as God or Spirit which always shows up only in hopeless dreams. Hopeless dreams are mere inarticulate or unplanned outcome of the brain, especially, when we sleep. Hopeless dreams appear when the brain is dormant and out of focus.

A dormant brain refuses to explore critically and remains unchallenged. Everything becomes impossible to a dormant brain except those things perceived by the individual's brain of what God has ordained. Faith in God becomes the only resort to those who fail to cope with realities of life. A dormant brain remains in a godly cage until it is brainwashed out of it or until it liberates itself to a life of self-introspection.

Dependence on one's community belief system without serious personal effort to accomplish a definite goal, always results in frustration, blame-shifting, deceit and inarticulate dependency on divine faith. *Mankind surely has many answers to many human problems today than it used to* during the pre-historic times. Human dependency on such icons as the Pharaohs, Caesars, Moses, Isis, Osiris, Dionysus, Mithras, Kama, Buddha, Krishna, Sikhs, Mazda, Yahweh, Jesus Christ, Mohammed, and hundred other Gods or God's messengers and divine leaders are no more relevant in contemporary world.

Most primitive religious persons of today can testify that life is more pleasant today than their predecessors'. This is so mainly because a few bold individuals had the courage to liberate their mindset or challenge the religious status quo and boldly face challenges of the real world. Religious leaders from prehistoric times to date have done everything possible to control people's mind with obsolete godly ideas and false hope, because deep down in their pastoral conscience they see that their personal livelihood depend on these religious practices. It is economic suicide for them to discourage the practice. This is a fact. You may call it the hidden purpose for encouraging religious faith. I guarantee a complete shut down if religious followers refuse to offer financial contributions to their faith leaders; none can survive for even one month.

'God' as a factor in human life has consistently proved fruitless, useless, and counterproductive. The 'omni' God or the all powerful, efficient, intelligent, benevolent, merciful, maker, destroyer, imperishable, and the 'always present' God are nothing other than 'a processed product of human brain cells.' They are not real but delusional.

The human brain, although, is *not yet* capable of processing *all* the acquired planetary information about how life first came into existence; if such a period as creation ever occurred, it will still be the human brain that would explore it. Invention of the microscope to probe the invisible world of micro-organisms, and telescopic satellites which probes the cosmic universe and beyond would not have been possible if some brave intelligent humans of our recent past had not paved a way at least for those alive today to understand the real coexistence of visible (macro-organic) and invisible (micro-organic) world. These are testable verifiable human phenomenon. Dreamers, revered magicians, idols, imposters, and masters of deceit, in the name of spirit or a cosmic deity called God, are gradually becoming extinct unrestrained. This is the modern trend because superstition and science are incompatible.

By their written scriptures, such as the Torah, the Holy Bible, and the Quran, there is one God, a single creator responsible for everything in the world. Each one has a different name for Him. God in the Torah is called Yahweh; The Holy Bible's God mutates Himself from Yahweh to Jesus Christ, to Son of God, to the Holy Ghost, to the Father, and finally God himself. In the Quran, the God is named simply as Allah. We are informed that all the three religions worship the same God who has Heaven as his residence. This god demands daily worship, praises, sacrifices, and specific performances from all followers according to their traditions. We can examine briefly some elements of these three faiths namely Judaism, Christianity, and Islam, and then apply common sense to their relevance, although followers are sadly forbidden to apply common sense on faith questions.

Judaism was and still is a Jewish religion. Christianity was born out of Judaism. Biblical Jesus Christ was a Jew from Nazareth,

Galilee. Although Jesus was born into Judaism his ideas were contrary to the Jewish cult's principles. Indeed there were dozens of his kind at the time he was alive. For instance, the name Jesus was a common Jewish name, and many people were called Jesus in most Jewish communities during Jesus Christ's lifetime. His followers were called 'the Christos' during his lifetime because his Jewish followers believed him to be the prophesised Christ according to Jewish traditional history. His followers were nicknamed as 'Christians' centuries after his death and acclaimed resurrection. Paul, Peter, and Barnabas are among the few first historical figures known to have initiated, internationalised, and amplified the gospel or the so-called '*good* news' of the deviant biblical Christ. He was a deviant because his religious principles deviated from the traditional Judaic ideas.

Jewish people who followed the laws of Abraham, Moses, and ancient Hebrew prophets, plus those who accepted the existence of a single God, called themselves Judaists. Ironically, it was the same Judaists who saw biblical 'Christ' as an imposter or one who claimed to be the prophesised Christ. He was not seen as the Old Testament's prophesised Christ. Jesus Christ's followers or admirers during his lifetime were *also* despised and seen as deviants by other Jews. Christianity as a religion therefore *began* as a phoney religion according to Judaism. Most Jews today believe the prophesed Christ is not the biblical Jesus. The Messiah is not born yet according to most Jews. They are still (the year 2011) waiting for *the birth* of the Christ.

It was this disagreement about Jesus as the Christ that brought about the fundamental difference or element among the Jews which eventually separated Jesus's followers from Judaism and consequently brought about a new religion two centuries later, known today as Christianity. Both Judaism and Christianity now thrive on the laws of Moses as prescribed by their cosmic god portrayed in the Holy Bible and the Jewish Bible—the Torah.

Islamic religion *emerged* into the religious map of the world about 500 years after the death of the biblical Jesus Christ. It was

Prophet Muhammad, a native of Mecca, who received spiritual revelation from Allah at Mount Hira in Mecca and who *transformed* the faith into a religion on 16 July 622. He made Islam a formal established religious institution.

Prophet Mohammad was an orphan at the age of six and an illiterate merchant at age twenty-five. We are informed that he received instructions or revelations from Allah (the same biblical God), through an angel, Gabriel, to serve mankind as the 'God-appointed' last prophet or Messiah. It is written that he lived to promote his message from Allah or God. Today, Prophet Mohamed has multitude of followers across the planet earth who recite Holy Quran verses as a means of worship and praise to Allah for His spiritual guidance. Islam, Christianity, and Judaism claim their roots from common ancestry—the God of Abraham, of Isaac, and of David, descending from Adam and Eve. The basic difference between the two religions is that Mohammed saw himself as a mortal human being. And also he was naturally conceived by his mother through sexual intercourse with his father. He lived and died, and was never resurrected. It was the male angel Gabriel who revealed himself in a dream to him and appointed him as the Messenger of God to carry Allah's message to mankind, according to Islam.

The biblical Christ on the other hand was conceived by a virgin mother who was impregnated by a male angel also called Gabriel. According to the Bible, Mary did not have sexual intercourse with Joseph (her husband) until after the birth of Jesus the Christ. Biblically, it was the virgin birth and resurrection in particular, which made Jesus a unique human being, *the Christ*, and the unique son of God among all God's children.

CHAPTER 3

The Holy Bible

i.) Introduction

It is very important to be aware throughout the following discussions that the object in question at this moment is a book, 'the Book' as the Bible was then called during and beyond Emperor Constantine's era in AD 228-337. The book, 'Holy Bible', is a man-made object, not a natural matter such as water and earth.

The Holy Bible is a bundle of papers or sheets on which are written words or recorded information *about Jewish people and their religious faith* only. The stories in the Holy Bible covers the period 1200 BC-AD 100—about 1,200 years before the birth of Jesus Christ up to about 100 years after his death. The concepts in the book were designed, planned, edited, and published by ordinary human beings. It was not written or documented or commanded and delivered magically by god from heaven. It was written neither by a single person or a group of people commanded by God or angles, nor via dream revelations.

None of the sixty-six authors ever knew or ever met Jesus in person when the man was alive. Everything in the book is hearsay. Historically, there are more than sixty-six authors, because the book of Deuteronomy alone is believed to contain eleven separate books written in different eras. Not a single word in the Holy Bible represents an eyewitness account. No letters or documents written by Jesus Christ himself or his disciples were left or found after his death, none whatsoever. This is a historical fact.

The original idea of creating a single document or a book which became known as the Holy Bible started about 300 years after the biblical Jesus Christ's death.

The *book*, the Holy Bible, is therefore 'not' and 'cannot' be taken as a sacred book. The holy title was given to the book to boost the divine or mystique image of the characters in the Bible.

The over three million words used to write the book solely represents the culture of a race of people, 'Jewish people', who lived in the ancient times in the Palestine peninsular area *as remembered* by Jesus Christ's followers over a hundred years after his death.

The book's global inferences during its initial setting in the ancient times was nil and essentially limited to Jewish world only. *Its original intended reading public was the Jewish church leadership only.* It is also very important to note that even at that time, after 900 years of Jesus Christ's death, very few Jews identified themselves as followers of Christ. Access to the Holy Bible by the rank and file and other cultures began after over a thousand years of Jesus Christ's death, or precisely, until the sixteenth. century, when a German, Martin Luther, started his Reformation crusade and made the Holy Bible available to ordinary folks.

The Bible as a document containing partial information about Christ followers was a guided book, specifically designed and written by Jews as a guiding text book for Jewish elderly church members only. At this point it is considered necessary to give a brief account of the main man responsible for the book's invention.

ii.) Irenaeus (Bishop of Lyons)

His actual date of birth is not known, but it is certain that he was born either between AD 115-125 and AD 130-142. He was certainly born about 142 years after Christ's death in a town called Smyrna in Asia Minor (modern-day Izmir, Turkey). Both of his parents were believed to be Greek Jews. He died as the Bishop of Lugdunum in Gaul (now Lyon, France) in AD 202.

He was an early-church Father and Apologist. He wrote extensively, especially about Christology, Mariology, and the codification, translation, and interpretations of the Gospels in the Christian world. It was Irenaeus who sponsored the collection of information about the fourfold Gospels of Matthew, Mark, Luke, and John. For instance, it was he who stated that the Gospel of John was written by John the Apostle, and Gospel of Luke was by Luke, who was the companion of Paul.

It was not until 1945 when the Gnostic scriptures were discovered in a small town called Nag Hammadi in Egypt, which revealed the existence of over twenty-seven *other hidden Gospels* dated back to Irenaeus' time. The choice of only four out of over one thousand witness texts incorporated in the New Testament of the Holy Bible was doubtlessly engineered by Irenaeus. Obviously, he was the central figure who formed the initial agenda of the founding fathers of Christianity during the second half of the second century.

Bishop Irenaeus central theological theme was unity of ideas and the teaching of the '*Good* News of God'. The *works* or the *deeds* of Jesus, his Apostles, and the attached families were the only parts considered important to let followers know, according to Irenaeus. The personalities or the biographies of biblical human beings were *deliberately cast out* because Irenaeus, and later Emperor Constantine I and the Nicene Creed, considered them irrelevant and heretic.

Again Irenaeus *opposed* Gnostic division of God into several divine 'Eons' or Apostles, such as the separation of church or the followers of Matthew or of Mark or Luke or John away from each other.

This was the prevalent practice at the time. There existed separate institutional churches of Mark, of Matthew, and the rest, each with similar Christology ideology and both competing for followers. He detested Gnosticism, a religion founded on personalities rather than on ideas of specific personalities. In reality, however, a religion which is based on half story or only the good part of a story alone; as opposed to the whole or both bad and good parts of a story of leading icons, opens itself to multiple suspicions. Irenaeus' decisions above posed several questions: If all the named characters in the Holy Bible were real humans, like you and I, why must their biographies not be told? Why must there be *two* or more separate Bibles for a single religion? The Gnostic Bible and the Synoptic Bible, and others share common faith based on the same biblical personalities, yet the two narratives are different and contradictory.

The Holy Bible is Synoptic, i.e. it is concerned with the works and deeds or philosophies of biblical people only. In contrast, *the Gnostic Bible mainly accounts for the overall biographies or detailed aspects* for the same biblical people. Irenaeus detested the idea of the Higher God and the lower or inferior God such as the Heavenly God and the earthly God (Christ) divisions. He maintained that Jesus manifested into Christ and into God himself; to form a perfect unity. The primary understanding of God's relationship with the universe as stated in the current Holy Bible was formatted by Irenaeus.

Among his numerous books was the five-volume book titled *The Detention and Overthrow of the So-called Gnosis.* In these books, he stressed the need for unity of the Old Testament and the Gospel; for example, *the sayings* of Jesus plus the spread of Apostle Paul's messages. In his best-known book, *Adversus Haereses* or *Against Heresies* written in AD 180, Irenaeus propagated the importance of avoiding Gnosticism or literal interpretations of the Gospels. Instead he instituted its divine application to life hereafter. He insisted that words attributed to Christ or God must not be understood in their literal meaning: but why so? If this person (Jesus) was really a human being, as all of us (born out of a woman's womb) and later transforms to become God, why Irenaeus should be worried if readers question his deeds and his human personal life?

He saw the idea of Christ being born out of Virgin Mary as necessarily, creating a totally new historical situation. The Virgin Mary became the mother of God (the Theotokos) as already discussed above. In the deepest sense of this word, Theotokos becomes God, which is a man-made entity. And indeed if Jesus' mother is the mother of God, are we all not Theotokos? Others saw Jesus becoming Christ after baptism by John the Baptist, or Jesus becoming Christ after his resurrection from death.

Irenaeus was a well-known Jew and a Greek scholar with distinction; he understood the definition of the Greek word '*Kyrios*', which means the 'Lord', very well. It was he who attached the title 'Lord' to Jesus Christ's name. The Nag Hammadi library as edited in the book called 'the Gnostic Bible' (see Bibliography) clearly shows that Irenaeus and his succeeding generations deliberately misguided Christians and the world at large since the mid-second century into believing in a cosmic God of his generation's creation. The sky was the domain of the spirits of the dead. And deep in the sky is the residence of our maker, God, who is endlessly watching the movement of everyone and everything under the sky.

If we can stop reading for a moment and try to recollect what we have individually heard about our particular cultural folklore or God, we would realise that there is virtually no difference in the so-called Heavenly God, or a God of any description, no matter the image we choose to characterise God. In the final analysis, God is simply the brain, no less and no more. Nag Hammadi's revelations will be dealt with elsewhere in this discourse. There we shall see the other side of the biblical coin.

The first *printed* copy of Holy Bible was made in the year AD 1417, i.e. 1450 years after the death of biblical Jesus Christ. By that year, the Book had been translated and transformed into several languages with dialectical differences and often with deliberate manipulation to fulfill individual faith's agendas. And indeed many alterations have taken place since the fifteenth century.

iii.) A Brief History

Before biblical Jesus Christ was born, the entire Palestine peninsula was a very small part of the Roman Empire. The seat of government of this mighty empire or country was in Rome, Italy. About 280 years after the death of the biblical Jesus Christ, Roman Empire had become too big and ungovernable by a single central body headed by one individual called the emperor working in concert with a council of senators. Several provinces were experiencing social unrest. Consequently, the empire began to collapse. It disintegrated into the East and the West empires. Each was headed by an autonomous emperor.

At first, Constantine I became emperor for the West from AD 312, or about 279 years after Jesus Christ's death: if we accept Jesus as a real human who lived for 33 years on planet earth. Religious organisations at the time were among the most effective organised civil groups besides the military in most regions or provinces of the empire. After a number of clashes between the West and the East for over twenty years, Constantine I was acclaimed the emperor for both West and East from AD 324 to 337. As a gesture of goodwill for the Christians' support, Constantine organised a council of bishops to search and to present the entire Jewish scriptures for codification.

In the year AD 325, a bishops' council known as the 'Council of Nicaea' reported their findings to Constantine I. After several months of serious intensive debate regarding such topics as 'who is who', classification of old and new texts, interpretation of Hebrew and Greek words and terms, presentation of ideas, choice of testimonies, which icon should be included or excluded or given a book title name, and their relevance to their ultimate religious agenda.

Emperor Constantine I agreed to preside over the first Ecumenical Council of the Christian Churches at *Nicaea* (present day city of Iznik in Turkey) 292 years after Jesus Christ's death. Please bear in mind that even up to that year there was no such single book known as the Holy Bible. This is a historical fact. The final document that was agreed on by this creed became *'the foundation book'* or what we know today as the Old Testament and the New Testament, popularly called the Holy Bible. Constantine declared Christianity as the official religion for Roman Empire during this period. It should be added, however, that many Councils were formed and a series of Conferences were held over the ensuing years to ratify the content of the Bishops reports which eventually formed the bases of the Holy Bible. About 8 of such major Conferences are summarized herewith.

Some of the major debates at Nicaea were *whether all* the over 1,426 names of men and over 120 names of women and their stories presented by the Bishops to the Council should be addressed *individually* or in *groups*, and if so how: which is more important than the other. Besides all that, the hottest debate at Nicaea was whether one God existed in *three persons* or in *one person*, i.e. Father, Son, and the Holy Ghost; should they exist in one individual person, 'Jesus'?

The Council finally declared the full divinity and full humanity of Jesus after lengthy violent debates between those who saw biblical Christ as human and those who saw him as divine. Bishop Irenaeus' written works of the second and the third centuries was one major source of reference for most parts of the reports. It should be noted that all the 300 members of the Council were concerned about the authenticity of their fellow members' research findings.

A second Bishops' Council (the Council of Constantinople, AD 381) was convened to clear some ambiguities about a Greek word '*homo-ousios*' in the texts which made Jesus' status as the '*Son* and the *Father*' appear to mean that the two are 'of the same substance'. It is noteworthy again that these intellectuals knew precisely the possible future implications of their deliberations. The *unnatural fusion of the Son and the Father* automatically makes the object, Jesus, a unique superhuman character worthy of all the wonders assigned to him.

Thus in 431 CE, the 'First Council of Ephesus' (the third Bishops' Council) was initiated to address Bishop Nestorius's (Nicaea Council member) views on Mariology, which again extended to embrace Christology as argued by Bishop Anastasius of Alexandria. For example, if Christ was both human and divine, who was Mary, his mother? Was Mary *'Theotokos'*, a woman who gives birth to God, if Jesus Christ is the same as God?

A fourth Bishops' Council, 'Council of Chalcedon' in AD 451 (twenty years after the third council) was established. The biblical Gospel of John's infusion of the word '*logos*' or 'word' and its pre-existence before Jesus was born, plus its cosmic implications created extensive debates and divisions among Christians at the time. Note again that even at this time (418 years after the death of Jesus Christ), the term 'Christians' was virtually applicable to an insignificant number of Roman citizens; although it was significantly gaining popularity in a few Jewish communities. Although Emperor Constantine I had decreed Christianity as the official religion for Roman Empire in AD 325, its practice in AD 450 was still predominantly Jewish. A great number of Jewish populations then, and even today, refused to accept Jesus' characterisation as *homo-ousios, Theotokos,* and *logos*' implications.

Thus the primary Christian theology developed into several branches thenceforth; hence Roman Catholicism, Eastern Orthodoxy, Anglicanism, Lutheranism, Eastern Christianity, Syrian Orthodoxy, Assyrian Churches, Coptic Orthodoxy, Ethiopian Orthodoxy, Armenian Apostolicism, and several types in the later years grew out of the initial cults.

A series of other Bishops' Councils followed after the fourth one; there were three major ones—AD 553, AD 680-1, Nicaea (11) AD 787. The central debate was about the definition of 'the Lord' '*Kyrie*' or '*Kyrios*.' Who was the 'Lord'—was it Christ or the Heavenly God?

In AD 851 (818 years after Christ death), essentially on questions about biblical Hebrew words translations, there was no common understanding for choice of words to narrate biblical stories. Hence another Bishops Council was convened in that year to choose the appropriate dubious expression befitting Christian theology. The Holy Bible as it is today has indeed passed through numerous theological grinding mills to achieve the desired perfect interpretations, versions, and scrutiny over the years to suit every generation's *religious agenda*. The Holy Bible we have today obviously represents today's understandings and interpretations of the Gospels as considered suitable by yesterday's Council of Bishops. This is a historical fact too. And it is the choice of this generation to carry the message forward unquestioningly or to modify the questionable parts that are incompatible with modernity. Christology may eventually render itself for burial in the graveyard of ancient philosophies, where many of its kind are in perpetual retirement.

A look at the past, therefore, shows that *every generation of Christians adjusts its faith to befit its needs, provided all the prophetic ambiguities are kept intact to enable flexibility of interpretations of the Jewish gospels.* It is noteworthy at this point to briefly explain the role played by the biblical architect himself, *Irenaeus,* who is often described as a bishop, a saint, a poet, a historian, a writer, and a librarian since 160 years after Jesus Christ's death. He was the leading theological architect who conducted comprehensive searches for ancient literary works of what might be called today as 'the biblical world literature'. The Christian theology of later years' could definitely not have been possible without his literary and philosophical contributions.

But, however, whoever, or whatever the book represents today, *the Holy Bible is essentially a book of two stories of Jewish people's past experiences—no more and no less.* The Old Testaments has thirty-nine *books*, beginning with Genesis and ending with Malachi. The thirty-nine books represent old stories before Jesus was born. The Old Testament contains *929 chapters*. It has *23,214 verses*, and *592,493 words*. It contains a total of *2,728,100 letters* only. It is important also to know that the first five books of the Old Testament are identical in words with the official Jewish Bible, the 'Torah'. Hence it is often referred to as the Pentateuch. The Bible contains real and imaginary Jewish history, mostly myths, prophesies of Jewish heroes before Christ was born, Jewish Gods, Jewish fundamental laws, and their code of ethics as perceived in Jewish culture before and after the birth of Christ. Everything about this book is about Jews as a cultural entity only. It has nothing to do with any other culture anywhere on planet Earth. The Apostle Paul and a few others attempted to universalise the works of Jesus several decades after Christ's death. Its contents may be similar with other cultures religious practices, but it will be a complete mistake to universalise anything in the Old and New Testaments as according to the book itself. The New Testament is so called because its contents represent new generation of Jews or Israelites who lived during Christ's Era (the century in which Jesus was born, up to the first 900 years after his death). The period covers his birth, his death, and his resurrection plus, of course, the time his admirers wrote his recollections. Again, it should be noted that this is essentially Jewish. Some books, chapters, and verses in the Holy Bible tend to universalise some Jewish private moral concepts to represent general public morality as if it is applicable to all cultures. The Bible specifically talks about the God of Israel and his beloved Israelites.

Some of the sixty-six books attempt to globalise some concepts; but it will be misleading and out of context to assimilate its content with other cultures' experiences. The New Testament is full of universal theological propagandas: although some contents may be compared to or interpreted to match a specific human situation or experience, it must not be forgotten that it is only similar to the situational experience. It is not an ordained godly precept. Everything therein is carefully crafted, man-made teaching or guiding principles of ancient Jewish theology.

The New Testament consists of twenty-seven *books*. Apart from the first five books, the remaining twenty-two books are generally believed to be letters written to Paul or by Paul himself. It begins with the book of Matthew and ends with the book of Revelation. It has *260 chapters* with *7,969 verses*. The New Testament contains *181,253 words*. There are *838,380 letters* of the English alphabet. The Testament contains the shortest verse in the Bible. It is John 11: 35, 'Jesus wept'. Thus there are *66 books in the Holy Bible*, with *1,189 chapters in all. Total number of verses* is *31,173. The total number of words* is *773,746*.

The number of letters of the English alphabet in the Bible is 3,566,480). The above statistics is from King James' version of the Holy Bible.

There are two chapters which are almost identical in the Holy Bible; these are 2 King 19 and *Isaiah 37*. There are *1,426 men* and *120 women* names mentioned in the Holy Bible.

The first 'general' question now is *who* the authors of these books were? The second is *when* the books were written? Third question is the *source* of the written information. The fourth is in what language or languages were they originally written? Were the original books translated into other languages? If so, by whom, when, and how was it done? Fifth question is who assembled and supervised the handwritten manuscript into sixty-six books. Why not sixty-seven or sixty-five books? Who is responsible for the numerous errors in the Bible—the ambiguities, repetitions, contradictions, missing verses, senseless statements, and incomplete sentences? Who accounts for the numerous geographical and chronological reference errors? Why are there several versions of the Holy Bible?

When we come to specifics of the individual books, thousands of legitimate questions pop up. In the Introduction of the book of Genesis this is what is found: 'The name Genesis means "origin." The book talks about the creation of the universe, the beginning of life, the origin of mankind, the beginning of sin and suffering in the world, and about God's ways of dealing with mankind, according to ancient Jews perspectives.' Explicitly, the biblical God is seen in the book as the god for all the 6.9 billion people on planet Earth today. The Bible completely disregards the existence of thousands of other cultural gods. The texts consider every other God as inferior. Our task at the moment is to search in the Bible, the legitimacy of this cosmic biblical implication.

The first eleven chapters of Genesis which deal with the creation of the world and the supposed early history of human beings, however, tell the story of Adam and Eve, Cain and Abel, Noah and the flood, and the Tower of Babylon to say the least. The first person on plant Earth according to Genesis is Adam (a male creature, followed by a female named Eve).

Chapters twelve to fifty continue with the history of the early ancestors of the Israelites. The first is Abraham, noted for his unflinching faith and obedience to the biblical God. He is followed with the story of his son Isaac and grandson Jacob, *who is also known as Israel*. Jacob's twelve sons are shown as the founders of the twelve tribes of Israel. Particular attention is given to one of Jacob's sons, Joseph, plus the events that brought Jacob and his other sons with their families to live in Egypt. It is significant to note that the main character of the book of Genesis is the Jewish God who does no wrong. He is one and the only entity whose power overrides all. All the questions raised above will be answered later with specific references to relevant chapters and verses written in the Holy Bible plus the actual evidential recorded history of the Holy Bible itself. But before that, a little bit of Roman Empire's religious history will be necessary to boost clearer understanding of modern-day Holy Bible texts.

Chapter 4

Christianity

i.) Roman Empire and Beyond

About 600 years before Jesus Christ was born, Rome, which is currently a city in Italy, was not just a city; Rome was a state, or in the modern sense, a country. It was a city-state. In those days, city states were very common in all the continents of the world.

Rome was declared a Republic in 509 BC when the Latin speaking Romans threw off Etruscan kings, which meant that Rome would no more be governed by kings or queens. The Republican government consisted of plebeians, *ordinary people*, but predominantly patricians, *rich people*, who served as senators. The history of the Roman Republic was one of continual territorial expansion.

By the year 275 BC, the Republic had expanded its territory beyond the borders of the city of Rome to include the entire Italian Peninsular. By 45 BC, the Republic had become a powerful nation militarily and had conquered the eastern and the western parts of Mediterranean and Gaul (France) under the leadership of Julius Caesar who made himself a military dictator and an imperialist leader.

Julius Caesar was assassinated in 44 BC by his own senators for being over ambitious. His nephew Octavian emerged as leader of the Republic after defeating Mark Anthony in the battle of Actium in 31 BC after Caesar's death. And this marks the turning point of Rome becoming *The Roman Empire* with *Gaius Julius Caesar Octavianus*, popularly called Octavian or Augustus, as its first emperor. *'Augustus' was a title given to the new Emperor. It* means *'Reverend'* in Latin.

After over twenty years, Augustus emerged as the first dynamic Emperor of Rome in 27 BC. Caesar's successor, Augustus, and all the succeeding Roman Emperors added 'Caesar' to their names, signifying their allegiance to him and his ideas. Augustus Caesar continued the expansionist policies of his uncle, Julius Caesar, until his death in AD 14. By this time most of the countries along the Mediterranean sea coast in North Africa and Adriatic Sea had been colonised and controlled by the Roman Empire. All citizens along the Palestine peninsula, Cyprus, Alexandria, Numidia, Mauretania (in Africa), and Spain, France, Armenia, and Germanic tribes across the River Rhine in Europe among others came under the umbrella of Roman jurisdiction. And very noteworthy in our present discourse of searching for our universal creator are the following relevant characteristics:

1. Roman imperial policies did not include indulgence in local politics and traditions of their territories.
2. Local government was carried out by locally appointed trusted citizens but supervised strictly by Rome.
3. Full economic exploitation of local resources for mutual benefit.
4. Internal revenue collection, taxes, excise, and import duties was of paramount importance.
5. Military defense against external and internal aggression was imperative.
6. Strict enforcement of Roman Laws in all territories to secure peace and justice.

7. All deceased Emperors were reincarnated as Gods, whose spirits were respected and worshiped as Gods in all acquired territories.

Julius Caesar was, therefore, the first God of the Roman Empire. The second God was Augustus Gaius Octavianus. At the birth of Jesus of Nazareth, the people of Galilee had temples already built for worshiping Julius Caesar as the imperial God. Augustus was initially addressed the Nephew of God because he was Julius Caesar's son.

Jesus Christ was fourteen years old when Emperor Augustus died. He was succeeded by Emperor Tiberius (AD 14-37) under whose administration Jesus Christ grew and practiced his belief. But the new emperor, Tiberius, projected Augustus as second God for the Roman Empire. Pontius Pilate was a prominent Galilean Jew. Hence his appointment as the governor of Galilee and Herod as the king of the province. Every community, however, had the right to practice its own religion or faith, as long as it did not interfere with the rights of others. But no God was superior to Caesar in the Roman Empire.

When Jesus Christ died at the age of thirty-three, Tiberius was still the Emperor and official God of Rome. The publicised revered nature of Jesus Christ in the Holy Bible would have shown up by all means in Roman historical archives if the biblical story was a real historical event which actually occurred. Indeed, the history of his mother and father and siblings, his miracles, his teaching, the plot against his life because of his anti-Jewish speeches, his claims as the true son of the heavenly God, plus all his inhuman characteristics would certainly have had a place in Emperor Tiberius's historical records. According to the Holy Bible, Jesus Christ was certainly an incredible human being such that his records, if indeed they were true human events, would not vanish from the official records of the Empire. When the eleventh Roman Emperor, Constantine I ordered a search for all Christological records, 292 years after Jesus Christ's death, the true written historical records must have been discovered and presented as evidence in support of the later years' biblical narratives before the Emperor would declare Jesus Christ's faith as the official faith of the greatest empire at the time. Tiberius may have recorded them, and Constantine must have known and possessed them. But as narrated by the author of the book of John in the Holy Bible, Jesus's *other records* must be concealed. No one can deny the natural reaction to this missing records mystery. It certainly opens the door to all logical speculations, and I sincerely hope that all Christian fanatics would henceforth understand the essence of this probe.

Now let us look at how the biblical story was received beyond 1,000 years after Christ' death, or precisely beyond the last Nicaea Bishops' meeting of 851AD, the years 'between' AD 1000-1600, the Middle Ages.

Before we begin, it must be made clear that by the end of the fourth Century, the seat of government of the Empire had been moved from Rome to Byzantium by Constantine, who renamed the city after himself as 'Constantinople'. In the meanwhile, the Emperor's declaration of Christianity as the official religion had caused the bishops to use Rome as its headquarters. The official departure from Rome strengthened the de facto religious role or power as well as the political vacuum position of the capital in the hands of the church leaders. The papacy had indirectly grown to become next in control of the city affairs. By the end of the fifth century, that is 292 years after the death of Jesus Christ, the imperial *universal* or the *catholic* religion popularly practiced by many was affirmed. *Roman universal religion* in the Empire became *Roman Catholic religion.*

By the beginning of the tenth century, that is by AD 900, *Christianity* and *Roman Catholicism* had become the same by definition. Through tax exemptions and several well-recorded religious, legal, political, and economic malpractices, Roman

Catholic as a religious institution became the richest and most powerful single institution in Europe. The Catholic religious machinery had revived the city's pre-republican glory—this time in the hands of the political clergy with the sward in one hand and the crucifix on the other. We are told today to believe that these and their later years' religious offspring are the Godly people who link us with the so-called Heavenly Father.

The imperial administrative headquarters had been removed back to Rome and fused together with religious Catholic administration responsible for all affairs of the huge empire. By AD 1400, Western Europe had been conquered and controlled by not just Rome but by *Rom*an Universal or Catholics. Thus Britain, Scandinavia, Austria, Hungry, Danes, France, Germany, Portugal, the Netherlands, Swiss, Slovenia, Turkey, and many more were not only governed by Roman Emperor but more so by Roman Catholic Church administrators in Rome. Catholicism spread rapidly systematically throughout these regions and their future overseas colonies as well. The interesting implication of the political holy alliance was that all political leaders within the dominion had to receive God's blessings from the headquarters at Rome. Caesar as the God of the Empire had been marginalised. Christian God of Israel, as perceived in Holy Bible, was *de jury* as well as *de facto* God for all; hence all kings and political leaders were required to obtain papal authority as well as certification to be recognised as rulers. In the meanwhile, the enormous size and the increasing lack of responsibilities on behalf of leaders of the empire had caused major cracks of the Empire's basic foundation. This contributed to its eventual downfall or disintegration. Imperial administration had obviously exceeded its optimum size.

Again as already indicated in other chapters, the orthodox religious leaders had been predominantly Jews. And by the early 16th century, local ethnic church leaders, who were not genealogically Jews but only had Jewish links, dominated the Christian leadership and fellowship. Christian theology, which was fundamentally a Jewish cultural philosophy, began spreading its theological virus to infest other ethnic cultural faiths throughout the Roman Empire. All Roman colonies were coerced to accept Christianity as the official religion to the detriment of locally cherished religions; at the same time, however, statutes of former Emperors—the Caesars as Gods, as well as the other Gods such as Mithras and Simons—were scattered in most provincial cities.

But, of course, this religious revolution did not take place without challenge by the then new generation of medieval religious intellectuals. By the end of the 17th century, legitimacy of Catholicism had been vigorously challenged by such thinkers as Bishop Athanasius of Alexandria, the Bishop of Hippo Regius in North Africa, Paula, the champion of women in religion, Elizabeth Schussler, and hundred others in Europe. Many European countries grew to detest autocratic rule of Romanic papacy and consequently declared unilateral religious and political independence. The birth of modern-day Christianity or its reformation, in effect, took place during this era.

Many things of course happened during this period about the Book, the Christians' Bible, such that if we are to tell all of them, perhaps *the whole world cannot contain the number of books that may be written.* I hope permission is granted by the author of Book of John for borrowing the quotation above. Among all the medieval Christians, none other than one person, Bishop Martin Luther, surpasses most legends and therefore need be analyzed here for his bold innovative ideas most of which made possible for you and I to have access to the Holy Bible today.

He was born on 10 November 1483 and died on 18. February 1546. He was a German, a theology professor of immaculate distinction, whose views about the then Orthodox Christianity, and in the opinions of many, changed the overall nature of

Christian religion. He was the medieval equivalent of the ancient Bishop Irenaeus and, indeed, comparable to the Apostle Paul of the first century Christian era.

His era is historically characterised as the age of Reformation, a notion he conceived and singlehandedly propagated with astonishing success.

In the year 1517, Martin Luther wrote and published a book titled *The 95 Theses*. In that document, he attacked corruption of the Catholic Church and the doctrines of the papal supremacy and clerical celibacy. He emphasised that it was 'the Bible' which served as the true Christian authority by itself. Interference of the Church in secular matters and the questionable unholy activities of the clergy, such as the sale of indulgences and antique holy relics, were irreligious. He insisted that the only justification of their being Christians was *'salvation by grace alone, through faith alone, for Christ's sake alone.'*

Martin Luther's revolt was popularly termed as the age of Reformation or Protestantism. According to him and his fellow protesters, *faith* is the key principle that underlies all godliness. His fourfold formula regarding biblical gospels are *solo christo—Christ; sola gratis—Grace; sola fide—Faith; sola scriptura—Scriptures*. Ideally, Luther's ideas were not different from the Chalcedon Creed of AD 451 (the singular image of Christ and God as exists in one person) already explained elsewhere in this discourse.

Briefly then, Martin Luther's crusade of the sixteenth century practically separated the protesters away from the Rome-based Universalist, or precisely 'Protestants' away from Catholics. From sixteenth century, therefore, several Christian denominations sprung up throughout Europe. Thus by the end of eighteenth, century (that is 1,667 years after the death of Christ), there had been more than a dozen different non-Catholic institutionalised Churches and mostly with their own versions of the Holy Bible. During his captive days at Wartburg Castle, Germany, between 1521 and 1522, Martin Luther translated the Greek version of the New Testament into German language in order to make Bible accessible to all people in the Holy Roman Empire of the German nation. In 1534, Martin Luther published his own version of the Holy Bible with his own denomination and fellowship in Germany. It was he who wrote, printed, and published the first non-Catholic Bible and made it the people's book in the churches, schools, and homes.

Before the 1540s, it was only the Pope, in person, who could decide what was good or bad for Christendom. Thus the overall interpretation of biblical scriptures was the sole prerogative of the Catholic Church under the leadership, guardianship, or the authority of Catholic Pope presiding at Rome. Only God could challenge Pope's religious powers then. But Martin Luther engineered a Protestant movement, which *decisively nullified* Pope's self-acclaimed religious authority forever. There are today hundreds of Lutheran Churches around the world. Martin Luther's birthday is still celebrated in certain parts of Germany.

In 1611, the king of England, King James I(who was at the same time also serving as the King of Scotland—King James V1) created or supervised a version of his own Holy Bible which is often referred to as 'the Authorised King James Version'. Needless to say, this era marks the birth of the Anglican Church, Methodist Church, Presbyterian Church plus of course several dozen others.

Now, dear reader, the main point of this lengthy expose is simply to state that the Bible you and I read today is essentially man-made and has nothing to do with holiness. It may contain spiritual stories written by wishful thinkers like you and I; that is all it is about. It is up to you and me to form our own individual opinions about its contents. There is absolutely nothing sacred or holy about the book; the 'holy' title attached to the book was deliberately chosen to mystify the ancient biblical fantasy. This sort of mystique is noticeably very common in every religion in every culture without exception.

ii.) Publication Dates of the 66 Books of the Holy Bible

Old Testament   New Testament

Books	Dates	Books	Dates
1. Genesis	1 BC	1. Matthew	60-85 AD
2. EXODUS	1 BC	2. MARK	60-70 AD
3. LEVITICUS	2 BC	3. LUKE	60-90 AD
4. NUMBERS	1 BC	4. JOHN	80-95 AD
5. DUETERONOMY	2-1 BC ★	5. ACTS	60-90 AD
6. JOSHUA	625 BC	6. ROMANS	57-58 AD
7. JUDGES	625 BC	7. 1 CORINTHIANS	57 AD
8. RUTH	6 BC	8. 2 CORINTHIANS	57 AD
9. 1 SAMUEL	625 BC	9. GALATIANS	45-55 AD
10. 2 SAMUEL	625 BC	10. EPHESIANS	65 AD
11. 1 KINGS	625 BC	11. PHILIPPIANS	57-82 AD
12. 2 KINGS	625 BC	12. COLOSSIANS	60+ AD
13. 1 CHRONICLES	4 BC	13. 1 THESSALONIANS	50 AD
14. 2 CHRONICLES	4 BC	14. HESSALONIANS	50 AD
15. EZRA	4 BC	15. 1 TIMOTHY	60-100 AD
16. NEHEMIAH	4 BC	16. 2 TIMOTHY	6-100 AD
17. ESTHER	4 BC	17. TITUS	60-100 AD
18. JOB	5 BC	18. PHILEMON	56 AD
19. PSALMS	950-586 BC	19. HEBREWS	80-90 AD
20. PROVERBS	6 BC	20. JAMES	50-200 AD
21. ECCLESIASTES	4 BC	21. 1 PETER	60-96 AD
22. SONG OF SONGS	950-200 BC	22. 2 PETER	60-130 AD
23. ISAIAH (3 bks.)	8-5 BC ★	23. 1 JOHN	90-100 AD
24. JEREMIAH	6 BC	24. 2 JOHN	90-100 AD
25. LAMENTATIONS	6 BC	25. 3 JOHN	90-100 AD
26. EZEKIEL	6 BC	26. JUDE	66-90 AD
27. DANIEL	165 BC	27. REVELATION	68-100 AD
28. HOSEA	8 BC	=======================	
29. JOEL?			

30.	AMOS	8 BC
31.	OBADIAH	6 BC
32.	JONAH	6 BC
33.	MICAH	6. BC
34.	NAHUM	8 BC
35.	HABAKKUK	6 BC
36.	ZEPHANIAH	7 BC
37.	HAGGAI	5 BC
38.	ZECHARIAH	5 BC
39.	MALACHI	5 BC

===========================

Source: 1) Encyclopedia of Ancient Egypt; by Margaret Bunson, Karen H. Jobs (2011)
2) Invitation to the Septuagint; by Moises Silva. (2001)
3) Ancient Records and Structures of Genesis: Wisemam, Wiseman, Nelson (1985)
4) The New Testament Documents by F.F. Bruce.

* Deuteronomy is a book of eleven (11) books. * Isaiah consists of three books.

Chapter 5

Who is the Biblical God?

i.) Book of Genesis (the Cosmic Creator God)

The Holy Bible contains several answers to this question. Firstly, the biblical God is a *He*. In the first chapter of the book of Genesis god is referred to as 'he', and so is the entire sixty-six books. Again the books are written in reported speech. The stories in the Holy Bible are being narrated by a reporter. There is nowhere in the Bible where the reader is told who the narrator is, and the source and the accuracy of information given.

For example, Genesis 1: 1 says *'In the beginning, when God created the universe . . .'* Someone or the author is about to tell the reader what happened during the time god was inventing the world. The first sentence in the Holy Bible is someone's report or a statement of God's action at a particular moment in time. *The sentence* is not formed as *'In the beginning when I, God, was creating the universe . . .'*

The question is where the writer was at that particular moment of creation. Was the writer present, or was he or she informed by another after the event? Was the information sourced from written documents or was it told by ancestors and finally got written down in a reported speech. Could the entire biblical story be taken as unquestioned truth without substantiating evidence of information source? Could it be wishful thinking, imagination, or indeed pure Jewish folklore?

The importance of the source question cannot be overlooked, simply because the object in question is a book, a man-made matter like anything made by man as already explained. We cannot eliminate human errors completely when someone is reporting an event which occurred at the time when she or he was not personally present. It is indeed reasonable to doubt authenticity of any story, including biblical stories, because none of the authors was alive when the actual events took place. Christians have nevertheless, accepted everything written in the Holy Bible as unquestionable truth. They have developed faith in the scriptures to the extent that unbelievers are often branded as evil.

There are indeed *too many* serious pertinent questions inherent in the Holy Bible that requires clear, concise, and unblemished answers. To authenticate the book as a holy document worthy of its acclaimed historical importance, it is crucial that this generation digs deep into its contents to either nullify it completely or glorify it once and for all.

Now, according to Genesis 1: 2-3, the earth was *'formless and desolate. The raging ocean that covered everything was engulfed in total darkness and the power of God was moving over the water.'*

According to this quotation, before God created the world or the universe there was the earth, ocean, water, and darkness, only that the earth was existing without light, form, or shape. In verse *14,* he commands, *'Let there be lights in the sky to separates day from night and to show the times when days, years, and "religious festivals" begin.'* At this point he had not even created humans so how could there be 'religious festivals'? There are no people to form a religious group.

From verses 1 to 25, God had not created humans yet; but in verse *26,* the author writes, *'then God said, And now we will make human beings: they will be like us, and resemble us.'* Surely indeed, we don't have to be professors in English language to understand the usage of such words as *we, like us,* and *resemble us* in the quotation. There is no doubt that at this point the author of Genesis believed the existence of other humans with god during or before he created the supposed first human being. 'We' and 'us' are used when referring to or addressing more than one audience. They are plural not singular words. At the moment of creation (even if such an event ever occurred) there was definitely someone present or that the reporter does not know what he or she is talking about. God was not alone when he allegedly commanded; or else the commanding action did not even take place at all. The obvious suspicion is that the author formulated the commanding words into god's mouth.

The male human being (Adam), who we are informed by the author as an invention of God in Genesis (even if it was true), could not have been the first male human. The report clearly shows other humans were present. Some Christians may naïvely claim that Jesus Christ was also present when God was inventing the first human, hence the use of such words as *'let us',' we', 'and resemble us.'* But we are informed that Christ had a mother, a female human mother. Who or where was the mother, Mary, at the time of creation? Was Jesus's mother, Mary, also there? The creation report in the book of Genesis would have mentioned it.

The quoted words above couldn't have come from anyone such as God who possessed all wisdom, if we accept these quotations in their literal sense. Perhaps there would be a clue somewhere in the Bible to justify misuse of words. There are certainly plenty of hints in the Old Testament, but what is more curious is the most vivid testimony in the fourth book of the New Testament where the Gospel according to John, written nearly a hundred years after the biblical Christ's death, and certainly millions of years after the supposed God's creation of humans, is saying that, *'Before the world was created, the "Word" already existed;* He *was with God and he was the same as God'* (John

1: 1). Verse 2 of the same chapter sums it up as follows: *'From the very beginning the word was with God.'* Some of the unadulterated Holy Bibles attempt to give definitions of some ambiguous biblical words. For instance, King James' version is saying that the Hebrew word 'logos' which was translated into the word 'word' also means 'man' or 'woman'. The apparent common sense here is that if the three words (man, word, woman) mean the same thing, and God intends to create a man it would be nonsensical to use 'word' instead of 'man'. Logically, the correct fitting word for the author could have been woman to avoid ambiguity, because creation or procreation starts in the womb. It is only the female species who possess the ability to conceive and to give birth. In effect, therefore, by implication, humans already existed before God came into the scene. Thus, the above quotation from John 1: 1 could and should have been written as 'Before the world was created, *man* already existed'. This indeed would make the quotation 'like us', 'we', 'and resemble us' more sensible. Thus, the authors of both Genesis and John are saying that even though God invented the universe plus the first human, other things and humans already existed. The two writers obviously intended to inform their readers that the biblical god was the original real creator of everything, but the authors failed to justify their claims. In fact they messed up.

Any open-minded critical reader of the first book of the Old Testament and the fourth book of the New Testament would conclude that their stories are pure allegories like Plato's allegory of the cave, St. Augustine's City of God, and William Shakespeare's dramas.

Biblical God's creation becomes more bizarre again when we turn to Genesis 2: 7: *'Then the Lord God took some soil from the ground and formed a man out of it; he breathed life-giving breath into his nostrils and the man began to live.'* First question—how old was this man on the very moment he was created. Was he a one-day-old child, a grown-up child, a matured man, or an old person? Was he twelve years old, thirty years old, or thirty-three years old like the fable biblical Christ? Most fundamental Christians would say this question is not important. But deep down in their conscience, their brain cannot delete this basic element of doubt.

If biblical God really made the first human, who made God himself? God couldn't have existed out of nothing. Someone somewhere must have formed him like he claims to have made the first person. Mankind is entitled to Gods' complete biography. Is God a Chinese, an Indian, a German, a black African, a Palestinian, a Jew, or an Israelite as it is claimed in the book of Deuteronomy chapter 32, 'the Song of Moses', *'He assigned to each nation a god, but Jacob's descendants he chose for himself'*; and in verse 39, *'I and I alone am God, no other god is real.'* I challenge the reader to read this entire chapter and judge who this biblical god really is. There is no way he could qualify to be the God of all! Impossible. He is phoney! He is a mythical Jewish God like any ethnic God found elsewhere, specifically, like the God of the readers' own culture.

According to Genesis 3:22, *'Then the Lord God said, Now [the] man has become(like one of us) and has knowledge of what is good and what is bad.'* In other words, [the] man, probably a Jew because everyone in the Bible is a Jew, has magically reached a matured age; and can reason like God (the Jewish God of Israel, because this God transformed later into the God of Israel); or indeed like any rational being. But to accept my translation and also apply *like one of us* clause in Genesis 3:22, as above, the author is clearly telling the reader of the pre-existence of humans prior to this fictional character God. 'Like one of us', like who and who or like God and who? Human earthly experience shows that becoming a man and possessing knowledge of what is good and what is bad does not happen by magical command. It is a process, and it takes years to grow into maturity. All human beings are born with zero minds, with no knowledge in the brain. If this mysterious creator really made man, as the author of the book of Genesis claims, every reader would certainly be convinced if supporting evidence was provided. For example, the first human's complete biography would have been sufficient, isn't it?

The Holy Bible's answer to the old question, 'How did the first human appear on planet Earth?' is absolute rubbish and childish. Human beings could not be a divine product. The biblical story of creation is a myth, and is theatrical. The roles of the actors of the drama are badly written, badly played, and senseless. Modern-day drama writer would do better than those ancient anonymous dramatists.

Another God's questionable creation is the female human being. According to Genesis 2: 21-23, 'Then the Lord God made *the man fall into a deep sleep*, and while he was sleeping, he took out one of the man's ribs and closed up the flesh.'

> He formed a woman out of the rib and brought her to him; then the man said, 'At last here is my own kind—a bone taken from my bone, and flesh from my flesh'. Woman is her name because she was taken out of man.

Read what Adam says: 'and a flesh from my flesh.' The author said, 'He formed a woman out of the rib and brought her to him'; no flesh is included. Right from the beginning of the so-called creation, several traces of misuse of words, illogical sentence formations, and a show of desperation to make this impossible act of God sound real becomes too obvious.

A modern rational being would imagine this god to be a trained physician conducting a surgical operation: God injects a massive dose of anesthesia; *the man falls into deep sleep.* Dr God, ready with his scalpel, *cuts and removes one of the man's ribs and closes up the flesh. He formed a woman . . .* Does this not remind us of the fiction most adults tell children—Dr Frankenstein story? Every culture universally can tell its unique kind of nurturing tales to children at nursery schools. As already mentioned above, the Bible is principally a Jewish document, and therefore the entire creation saga is a Jewish folklore which has no relevance in reality to any culture whatsoever. In forming the first woman for instance, notice that God used only the single male rib with no other substance, such as soil in the man's case, to create the first female. The Bible does not mention use of soil or any material other than a rib. It is a fact that every adult human being possess 24 ribs, but in this case, the Jewish God-made woman has only one rib and the God-made man therefore ended up with only 23 ribs after the surgery. This means that the true descendants of these two first humans on planet Earth (every woman and every man), you and I, would necessarily have identical anatomy; one rib for every female and twenty-three for every male if all of us are descendants of Adam and Eve. The couple's descendants are Jews, of course. Hence all Jews are supposed to possess same skeletal composition; but Jews happen to be humans too, and Jewish males and females do have 24 ribs each just like any other human being. Where are Adam and Eve's descendants? Have they perished? Most people would be tempted to put words into the author's mouth or biblical God's mouth that soil was also used to form Eve. We have to be very careful when reading the Bible.

Once upon a time, a three-year-old boy asked his mother where children come from, and mom said, they come from heaven. The child asked again, 'Mom, where is heaven?' Mom answered, 'Heaven is in the hospital.' He asked again, 'Is that where I came from?' This time mom got angry and told the kid to shut up.

We have passed this kindergarten-age mentality of the ancient folks, and surely adult rationality should permeate our judgement of how god could have formed a human being—Adam with just a soil, and a woman with a single male rib. Evidently, many words are put in this fictional God's mouth by the authors of the entire sixty-six books of the Holy Bible, but in this particular chapter, the author goofed.

If this God is really incredible as He is portrayed in the Bible, why did He not simply just command man and woman to appear from nowhere as He is supposed to have done with other animal races? Besides, He is supposed to be the omniscient and omnipotent. Nothing is supposed to be impossible to Him. This whole idea is childish, foolish, primitive, and not acceptable.

Now let us examine Genesis 3. This section deals with the disobedience of man, God's pronouncements or judgments and punishments, having already put the first married couple (Adam and Eve or *Hebrew translation* 'Mankind' and 'Life') in the Garden of Eden. God, according to the Holy Bible, knows everything everywhere at all times. Everything is preplanned by him. If indeed this is true, he would definitely have known in advance what could have happened in the Garden of Eden before he placed Adam and Eve there, because it is said that *he* was pleased with his creation. God had no reason to test man or in fact conduct any test of any kind. The fable of the snake, the tree of knowledge, and the whole episode of man's disobedience to God are pretty childish and an insult to human intelligence.

Adam, according to Genesis 2: 21-23 (quoted above), had already expressed a degree of intelligence when he said to God that 'at last I have one of my own kind' and expressed a need to give Eve a name. God is either unintelligent for not knowing or anticipating in advance what would have happened in the Garden, or the author does not know what the hell the creation story is about.

In Genesis 3: 1-5, the sneaky snake is even wiser than the God-made humans in the Garden of Eden. *'The snake replied, that is not true, you will not die. God said that because he knows that when you eat it, you will be like God, and know what is good and what is bad.'*

In Genesis 3: 20, *'Adam named his wife Eve, because she was the mother of all human beings.* There was no other human being then anyway, so what was Adam talking about anyway!

In verse 21, *the Lord God made clothes out of animal skins for Adam and his wife, and he clothed them.'* In Hebrew language Eve means Life, and Adam means Mankind. Adam, a lonely freshman on planet Earth, somehow has the knowledge of what a woman is, plus a need for her name. Adam calls Eve no other name than 'life'; 'The mother of all human beings'? Who is the real mother of all human beings—God or Eve or our own mothers? Is it not true that mothers are 'really' mothers of all human beings as verse 20 is clearly telling us? We are already informed that God is our heavenly father, but we are not told who our heavenly mother is.

Genesis 3: 1-12 is about disobedience of Man to God, his Maker; and the sole reason why millions of years later Jesus Christ was born and had to be crucified for.

> Now the snake was the most cunning animal that the Lord had made. The snake asked the woman 'did God really tell you not to eat fruit from any tree in the garden?' 'We may eat the fruit of any tree in the garden, the woman answered, except the tree in the middle of it. God told us not to eat the fruit of that tree or even touch it; if we do, we will die.' The snake replied that is not true; you will not die. God said that because he knows that when you eat it you will be like God, and know what is good and what is bad. The woman saw how beautiful the tree was and how good its fruits would be to eat, and she thought how wonderful it would be to become wise. So she took some of the fruit and ate it. She gave some to her husband and he also ate it. As soon as they had eaten it, they were given understanding, and realized that they were naked; so they sewed fig leaves together and covered themselves. That evening they heard the Lord God walking in the garden, and they hid from him among the trees. But the Lord God called out to the man. Where are you? He answered, I heard you in the garden; I was afraid and hid from you, because I was naked. Who told you that you were naked? God asked. Did you eat the fruit that I told you not to eat? The man answered. The woman you put here with me gave me the fruit, and I ate it . . .

All the above quoted verses are useless, useless, and useless; it is self-evident in the scriptures above that the woman already possessed the knowledge of good and bad. for instance, '*she thought how wonderful it would be to become wise.*' This word *thought* is intellectual exercise duly processed by her brain with pre-existence of information about wonderful things and what wisdom means. This is natural with all humans and animals. It certainly has nothing to do with god or anything mystical.

And if Genesis 3: 20 is saying, '*Eve is the mother of all human beings,*' how can we take the author seriously when by definition 'mother' means origin or source, a title befitting God only? Is this not the natural truth already known to all human beings, that our mothers are our creators, that it is impossible for anyone beyond our mothers to be our maker? Throughout human history, all humans are known to originate from the woman's womb after sexual intercourse. And evidently indeed Genesis 3: 22 is not ignoring this universal obvious truth about *Mother-nature*; yet Christians tend to ignore this and rather accept Genesis 2: 7—God used only soil to form the first man—and Genesis 2: 23—only one male rib to form the first woman. Too much is at stake here! Human rationality is that which make human beings different from other animals. The Bible appears to be successful in forbidding readers from rationalizing biblical texts. Rationally speaking, Genesis doesn't make sense.

The missing biography of this odd couple does not astonish Christians either because it is accepted as the gospel truth; hence, for instance, the couple's ages at the creation moment need not be known. Christians consider this pertinent question as heretic and irrelevant. Since the biblical God pronounced them as husband and wife, it must be presumed that both were consenting adults when God created them.

In all known cultures, however, people of the same blood, like Adam and Eve, are seen as brother and sister, not as married couple. But in his own incredible wisdom, Genesis' God makes the first male and the first female—people of *the same blood*—spouses. The same god condemns sexual intercourse among brother and sister in the book of Deuteronomy 27: 22, *'God [curses] on anyone who has sexual intercourse with his sister or half sister.'*

These contradictions would not be a big deal if we accept biblical story as mere fiction or mere artistic drama, but if we take such statements as true words of our maker, then there is a problem. How can we expect God to indulge in profanity, *to curse*? He is supposed to be beyond common rationality. Can God really curse, or damn? This is gross and revolting.

According Genesis 4: 1, *'Adam had intercourse with his wife, and she became pregnant. She bore a son and said. By the Lord's help I have acquired a son . . .'*

Who taught Adam and Eve what sex meant? Neither of the two had a mother, brother, or sister, only their maker, God, (who was also a father and mother), whose fatherhood was indeed an abysmal failure. If god can really make a human being with soil, why would there be a need for sexual intercourse between Adam and Eve? Would it be sex for fun or sex for procreation?

No responsible father would tolerate sexual affairs among his own children. I am not sure whether the author of Genesis did intend to promote incest, or whether its implications were not anticipated. For what purpose did God create vagina, uterus, womb, ovulation, menstruation, gestation, pregnancy, and penis, erection, testicles, sperms and the rest? If God actually created these in humans, why were they made? By his commandments, all these could have been made to appear magically. In fact the argument for God's existence really defeats its self absolutely. The architect of this biblical drama of creation and human ancestry failed to account for these intricate details. This failure makes the biblical God's role baseless and baloney.

Genesis 6: 5-7 read as follows:

> When the Lord saw how wicked everyone on earth was and how evil their thoughts were all the time, he

was sorry that he had ever made them and put them on the earth. He was so filled with regret that he said, 'I will wipe out these people I have created, and also the animals and the birds, because I am sorry that I made any of them.

Holy God! A faultless, invincible originator of all things and all humans, the one who knows yesterday, today, tomorrow, and a thousand years to come; how could he possibly not have seen this wickedness of man before designing him? Who is he kidding? Is God saying humans must not use the brain, which he is supposed to have given us? What is the use of the brain then? If human beings can make God, the Almighty, feel sorry and fill with regret, humans must be congratulated for being smarter than God. This god must be a nutcase!

Humans have certainly made God, the biblical Genesis God, very stupid indeed. The above quotation is truly ungodly and deserves no space in the so-called Holy Bible.

By its literal definition already cited elsewhere, God, apart from the obvious disqualification of god's title, is very irritating and naive. How God could not be so intelligent to foresee this is unbelievable. Or should God be so remorseful and absentminded?

The author has failed to justify the possibility of god's existence through God's acclaimed deeds. There are too many loopholes in the allegations. God is an impossibility. There can never be a God. God as a word can only exist if you, the reader, condition your mind that there *must* be a God regardless of what anybody says.

God is simply a condition of the human mind. It is never a dependable entity with divine powers. A purely delusional therapeutic notion essentially based on false precept which breeds false hope. *Faith of the 'self' and nothing beyond the self* should be the ideal thing to maintain in place of faith in a divine God. This notion may be convincing because it is self-motivated and results can vividly be identified, is foolproof, self-evident, and experienced individually.

The genesis of the so-called Holy book is grossly loosely created. If there is such a single founding father for the entire universe at all, which I consider as void, it cannot be the God as explained or written in the Holy Bible, or specifically, in the book of Genesis. No way!

In Genesis 32: 30, Jacob claims, '*I have seen God face to face, and I am still alive.*' It is also claimed elsewhere that no one can see God's face and stay alive. Which is which?

Again, similar ungodly behaviors are written in Genesis 19. The bizarre story of *Sodom and Gomorrah where God lost control and burned the twin city with burning sulfuric acid* is awfully sinful and ungodly. Can God, the managing director of our universe, really lose control? Who will be in-charge of this complex world, if this acclaimed super intelligent entity can behave like you and I?

We read again of Lot's wife who turned into a pillar of salt after which his two daughters, for lack of sex and procreation, intoxicated their natural father and subsequently had sexual intercourse with him. '*They called out to Lot and asked where are the men who came to stay with you tonight? Bring them out to us! The men of Sodom wanted to have sex with them*'. Endorsement of homosexuality and sickly perversions of all sorts are written therein for no logical purpose other than sheer dramatisation of biblical God's infamous supremacy. The author of the book of Genesis concludes with the death of Joseph in Egypt. The book also sets up a stage for another fictional character called Moses, who initiates messianic duties throughout the ensuing four books of the Old Testament, namely Exodus, Leviticus, Numbers, and Deuteronomy. The first five books of the Holy Bible are known to be exact copies of the famous Hebrew Bible 'the Torah'.

These four books appear to contain a continuous fabricated history from the time of Adam and Eve through Cain and Abel, Seth, Enoch, Noah, Abraham, Isaac, Jacob, David, et al. to the death of Joseph in Egypt. It is interesting to note that there is no credible historical evidence to support any of the stories narrated in book of Genesis. None whatsoever has been accepted, both archeologically and historically by several eminent scholars. (Refer bibliography.) Hundreds of names of people and places mentioned in the book have been searched by hundreds of experts for clues in support of the authors' claims, but none is flawless. Most critical readers accept Genesis as a classic ancient Jewish novel. It is certainly not a genuine place to look for a cosmic designer or creator popularly called God. No reputable modern university would accept the hypothesis presented in the Bible for accreditation. No, it is not the place. You cannot find genuine God anywhere, not even in the Holy Bible. The only thing apparent in the Bible is mythology, theatrical drama, dreams, and magic.

No wonder the second millennium clergy prevented ordinary church members from reading the Holy Book. It is still not a book worthy of reading with reason, particularly the first eleven chapters of Genesis which deals with creation of the world and history of God's early connection with mankind. The remaining thirty-nine chapters set a stage and transfer the Jews' god's powers to humanity, implying that Jews were the first humans God created to inhabit this planet. Thus, over the centuries, their ancestors spread across the whole planet. As the Jewish God is acceptable to Jews and their cohorts, so must other Gods be deemed acceptable to all other ethnicities. A single universal God is wishful thinking because its hypothesis is doomed at its base. The idea is void because it can never have indisputable supporting evidence. It will forever remain a cliché or a money-making tool for smart guys in all human cultures. There is not the slightest convincing reason why there must be a single universal creator of nature. Why? Is this world not too complex for a single engineer to invent? No matter how brilliant or mysterious that entity might be construed, its possibility can only be pictured in the mind if the brain is trapped with the Genesis concept of creation. To think of the existence of a possible cosmic inventor is a waste of time. God does not and cannot exist. Book of Genesis failed to prove God's existence.

Biblical writers have gone to the farthest extent to deliberately mold a human being—Jesus Christ—to substitute the Jewish God with such ambiguous Hebrew and Greek words as *logos* (word), *Kyrios* (Lord), *Theotokos* (one who gives birth to God—Mary), in primitive days. It did not work then, even with intimidation, and it cannot work now with persuasion. The universe is understood differently today. It is no more a dome as perceived in Genesis 1: 6, with localised horizon representing the overall size of the globe. It has over 6.9 billion brains occupying vast areas of the planet and determined to probe the planet as deep as possible to make life worth living, and less and less dependent on non-verifiable dogmas and primitive science like God and his cosmic creation.

Admittedly, only a fraction of world population is willing to question dogmas relating to the source of man, such as the creator God in Genesis. It is hoped that this fraction will continue to increase and eventually wipe out these man-made precepts for ever. But before we turn off the book of Genesis, let us flip over quickly to the twenty-fourth book of the Old Testament—Jeremiah 23: 23-24: '*I am [a] God who is everywhere and not in one place only. No one can hide where I cannot see him. Do you know that I am everywhere in heaven and on earth?*' The key word in this quotation is 'a', which in English language represents a function word used before singular nouns when the referent is more than one. There is a difference between 'a God' and 'the God'.

In this case, 'a God' means that there are other Gods. Jeremiah's god has an assignment to be everywhere in heaven and on earth. There must be *the* God who assigns duties to *a* God. Jeremiah's god is like a watchman God, who is controlled or directed by his superior god to be *everywhere* on earth and in heaven, just as there is a God of Israel for Jews only. Genesis' God is a creator or a designer of earthly things, and the New Testament God, Jesus Christ, is a redeemer God. It will be misleading to interpret Jeremiah's God as the cosmic God. Jeremiah's God is as fictional as the other biblical characters.

ii.) Book of Exodus (God of Israel)

The author's name is anonymous. Exodus is a Greek word which means 'departure' or 'going out'. It refers to the central event in Israel's history, which is described in this book as the departure of the people of Israel from Egypt, where they had been slaves. The book has three main parts: first, the freeing of the Hebrews from slavery and their journey to Mount Sinai. The second is God's covenant with *his people* (*Jews*) at Sinai. It gave them moral, civil, and religious laws to live by. The third is the building and furnishing of a place of worship for Israel, laws regarding the priests, and worship of the God of Israel. Above all, this book describes what the Jewish God did, as he liberated his enslaved people and formed them into a nation with hope for their future. The lead actor of the book of Exodus is Moses, the one whom God chose to lead his people, the supposed enslaved Jews, from Egypt to God's Promised Land. The widely known part of the book is the Ten Commandments in chapter 20.

A slave is someone held in bondage and maintained as a chattel, or considered as household equipment by another. A slave has no independent mind. A slave's brain is capable of taking instructions and not to give instructions. A slave, therefore, has an owner, like a cow with an owner. Human rights of a slave is owned by another.

Egypt is the name of a country in the north-eastern part of Africa. The country has been in the same geographical area or map for thousands of years. It is among the few countries in the world whose name has remained intact for thousands of years. And there has never been a time in Egyptian history when *all* Jews suddenly departed Egypt. There is zero recorded history.

There has always been a Jewish community in Egypt for millions of years. There are Jews today, in 2011, the many descendants of Jacob, who have had no country of origin other than Egypt.

The Holy Bible, especially the book of '*Going out*' or *Exodus*, has mentioned Egypt a number of times, because it presents a pivotal role, or indeed a highly significant position for the main actor, Moses, and Jewish residents in the ensuing books.

For instance, in Exodus 1,

> The sons of Jacob who went to Egypt with him, each with his family, were Reuben, Simeon, Levi, Judah, Issachar, Zebulun, Benjamin, Dan, Naphtali, Gad, and Asher. The total number of these people who directly descended from Jacob was seventy. Jacob's son Joseph was already in Egypt. Note this; In the course of time, Joseph, his brothers, and all the rest of that generation died, but their descendants, the Israelites, had many children and became so numerous and strong that Egypt was filled with them. Then a new king, who knew nothing about Joseph, came to power in Egypt. He said to his people, These Israelites are so numerous and strong that they are a threat to us. In case of war they might join our enemies in order to fight against us and might escape the country. We must find some ways to keep them from becoming even more numerous. So the Egyptians put slave-drive over them to crush their spirits with hard labor.

The king later ordered midwives to kill all Hebrew male babies and let the girls live. When this did not work, he ordered the Egyptians to look for all new born Hebrew baby boys and throw them into river Nile. According to the author, these affairs went on for several generations, about 430 years.

At this point, it is observed that a stage had been set for the next lead actor to link the past, present, and the future of biblical God's works; Moses is henceforth invented as God's mediator.

A couple, descendants of the Levi family, had a baby boy they feared to keep, because of the king's commandment to kill all Hebrew male babies. Note here that Levi is one of Jacob's twelve sons who chose to make Egypt their home. The baby is put in a basket made of reed and tar and given to the child's sister.

Exodus 2: 5 says, *The king's daughter came down to the river to bathe, . . . suddenly she noticed the basket in the tall grass and sent a slave girl to get it.*

The princess saw a baby boy crying in the basket. This is obviously one of the Hebrew male babies, she said. She took him home, gave him a name 'Moses', meaning' pull out' in Hebrew, placed him under someone's care, until the baby became a matured man. Moses eventually found himself in King Jethro's palace, where he accepted to live and married a princess called Zipporah who bore him a son called Gresham. According to verses 23-25,

> Years later the king of Egypt died, but the Israelites were still groaning under their slavery and cried out for help. Their cry went up to God, who heard their groaning and remembered his covenant with Abraham, Isaac, and Jacob. He saw the slavery of the Israelites and was concerned for them.

The interesting element of biblical authors is their lack of consistency and awareness of who the god in the Holy Bible really is supposed to be. We are informed by the author of the book of Genesis that the biblical God is ever present, everywhere, knows all at all times, and most powerful, and above everyone and everything. The God is ageless and never dies. This means at the time of the so-called Exodus, mankind had been on planet Earth several millions of years with god, and this god had been silent and inactive throughout until the death of an Egyptian king that *God suddenly remembered* the pact he signed with Abraham, Isaac, and Jacob. God suddenly recollects that his chosen people are suffering and needs a god-chosen leader to liberate them. Where was Adam and Eve's god or the Hebrew god or the god of Israel? What is god's reason for waiting 430 years before he chose Moses to liberate his chosen people, if this story is true? Can God suffer from amnesia? Of course not, but what is the big deal if the story in question—memory loss—is a theatrical drama.

The Israelites had been in Egypt for 430 years (Exodus 12) of constant cruel enslavement according to this book. In Genesis 15: 13, the length of time is stated as 400 years, while verse 16 gives three generations. Many historical data in the Bible are grossly inconsistent, ambiguous, and vague. If this does not confirm the fictional aspect of the biblical God story, then I don't know how anyone can conclude otherwise.

If the biblical God can be absent-minded, then his brain must be examined. He becomes as fallible as Adam and Eve, and indeed like you and I. He loses all his infamous 'Omni' titles. According to Exodus 3: 1-22, '*God revealed himself to Moses*' when Moses was then a shepherd living in Remises, Egypt. Yet Moses managed to travel about 900 miles to Mt. Sinai to receive God's instructions. '*There the angel of the Lord appeared to him as a flame coming from the middle of a bush. The bush was on fire but that it was not burning up.*' This reminds us of Saul and his companions' trip to Damascus in the book of Acts 9: 3 written in the New Testament about two thousand years after Moses' event.

> ...suddenly a light from the sky flashed around him ...
> He fell to the ground and heard a voice saying to him
> Saul, Saul! Why do you persecute me? Saul asked; who
> are you? I am Jesus whom you persecuted. The voice

said. The men who were travelling with Saul heard the voice but did not see the light.

In this case, Moses went closer and heard God's voice calling his name.

'I am the Lord of your ancestors, the God of Abraham, Isaac, and Jacob. I have seen how my people are being treated in Egypt, their cry for rescue, and all their sufferings' (Exodus 3: 6).

'Now I am sending you to the king of Egypt so that you can lead my people out of his country to a spacious land, one which is rich and fertile; land of the Canaanites . . .' (Exodus 3: 10)

What we are reading here is the genesis of Adolf Hitler's *Lebensraum*—the land acquisition policies of a German political leader between 1933-45. *Lebensraum is discussed in the next chapter.* 'Doubting Moses' asked God his name; In verse 14, '*God said, I am who I am. The one who is called I am.*' Note that the Hebrew word which gives this translation is Yahweh or Jehovah. Henceforth the biblical God who exists in the minds of humans has pronounced his name as 'I am'. In all fictions every name is OK; hence we need not question it.

For thousands of years, generations upon generations have come and died; the supposed designer of our universe finally talks to a Jewish ancestor, Moses, what his highness's name really is, the honorable Dr 'I am'.

Again if we travel ahead of time, another two thousand years to the New Testament, the book of John 1: 1-4, '*Before the world was created, the Word already existed; he was with God and he was the same as God'. From the very beginning the word was with God. Through him god made all things; not one thing in all creation was made without him. The word was the source of all life.*'

The other name of our creator or our universe's designer, according to the author of the book of John, is 'Word'. His first name may be stated as; 'I am' and his last name, 'Word'. God's full name as of now is Dr I am *word*. Remember that he was the surgeon who put Adam to sleep, cut his chest open, and removed a rip. He deserves the 'Dr' title. Sounds great! In Genesis and Exodus, however, his traditional names were *Elohim, Yahweh, El, and Isra-el* before Christ's era. In all dramas, any actor can be given any befitting name; hence the script writer of the book of John cannot be criticized for using 'Word' as our creator's name. If, however, we go to the source, we will discover that the Hebrew word *'logos'*, which was translated into 'word' in English, also means *'woman'* or *'man'* in Hebrew. If we accept the plausibility that the entire sixty-six books of the Bible was initially written in Hebrew before being decoded into other languages, why didn't the author of John choose to use the word 'woman' or 'man' instead of word. Again a masculine pronoun 'He' is used in place of 'Word' instead of 'It'. By implication, however, if the author had used man or woman, it would have meant his recognition of the pre-existence of humans in the universe before creation, be it a man or woman, before the so-called god initiated whatever he claims to have invented. Obviously, the choice of 'Word' was deliberate. It established a curious stage for the intended fictional character 'the living god', 'Christ', to assume the role of the already existing invented Jewish god as above mentioned in the first five books of the Bible. Continuity of the 'gospels' novel is definitely essential. The use of man or woman in place of word would invalidate god as the originator of the universe. The role of this character, Jesus the 'Christ', will become apparent when we reach the New Testament.

The book of Genesis and book of Exodus are deliberate preparation ground for the Human God's—Christ—legitimacy in the New Testament. However, it is obvious at this point from *Genesis 1: 26, John 1: 1-4,* and several biblical chapters and verses that the biblical God could not be the creator of anything in reality.

The universe, humans, and all matter were already in existence before this character god was projected into the global scene.

For those who still believe that there must be a single designer of the world, they must either re-read the Holy Bible with open conscience or that they should look somewhere other than the Holy Bible for that single god who did this creation. They can also ask themselves the question *why must there be a single creator and not multiple creators,* or consider the most plausible scenario as proposed by Charles Darwin and other evolutionary theorists. Scientific propositions are testable, verifiable, and always self-evident. The genesis of you and I is scientifically demonstrable with the union of ovum and sperm to produce zygote, which in turn passes through several changes before the final human being is formed.

It is also a known fact that no God or Spirit has ever been found inside the female genital system who designs the process or influences the fetus. By scientific means, the final human being can be seen with the naked eyes and even operated on if a defect is detected inside the womb before the baby is delivered. Where comes in this delusional miracle God?

Why can't human beings see their mothers and fathers as their makers? Why? Why must there be the first man and the first woman plus the first in every organic matter. Why can't we accept that a single designer of the billions of organisms on planet Earth is wishful thinking, a delusion, and an absolute impossibility even to the biblical god himself? Why? I challenge every religious person to re-read their scriptures completely and carefully, and I bet their ultimate conclusion would be no less than sheer faith or their mindset which has engulfed or trapped them in their religious worship and addictions.

The author of the book of Hebrew in the Holy Bible explicitly narrates the importance of faith in the acceptance of god as the creator. How about faith of the inner self rather than the outer self? For instance, *Hebrew 1: 1-3,* says, *'To have faith is to be sure of the things we hope for, to be certain of the things we cannot see. It was by their faith that people of ancient times won God's approval. It is by faith that we understand that the universe was created by God's*

word, so that what can be seen was made out of what cannot be seen.'
Please pause for a moment and read the last sentence again. It is surprising why the Bishops' Councils of fourth and fifth centuries permitted the inclusion of this sentence. The book of Hebrew contains only thirteen chapters and the author is informing readers that the whole script is a letter or epistle, probably, written by Paul but definitely addressed to skeptic Hebrews. Paul is loudly amplifying the essence of faith, that with confidence and trust nothing is impossible. Considering the time he was writing (approximately about thirty years after biblical Christ's death, and several millions of years after the supposed biblical God's creation) humanity has depended on superstitions, magic, myths, and institutional faith systems.

Knowledge of our environment was virtually zero then, compared to modern times where such phenomenon as computers, Internet, genetic engineering, stem cell research, cellular phones, and thousands of inventions have revealed multiple source secrets to mankind. These revelations were hitherto God's creations. And certainly, if Paul was alive today, the theme of his gospel would not be different from our present discourse. He would surely preach faith in inner-self of mankind than the outer-self or divine faith. He would tell his audience that in the past he preached about faith, hope, love, and knowledge as the basic element of God's relationship with mankind; but in the twenty-first century world, these elements reside in the individual human self. Have faith in yourself, acquire quality knowledge of a specific faculty, maintain hope and trust that you can do it, love what you do best and with zero reliance on god, and you will no doubt enjoy your own fruits.

The book of Exodus is concluded with a complete set up of a place of worship by Moses as dictated by the Jewish God. For instance, in Exodus *39 32-43,* All the work on the Tent of the Lord's presence was finally completed. The Israelites made everything just as the Lord had commanded Moses. They brought to Moses the Tent and all its equipments, its hooks, its frames, its crossbars, its posts, and its bases; the covering of rams' skin dyed red; the covering of fine leather; the protective curtain; the Covenant Box containing the stone tablets, its poles and its lid; the table and all its equipments, and the bread offered to God; the lamp-stand of pure gold, its lamps, all its equipment, and the oil for the lamps; the gold vary; the anointing oil; the sweet smelling incense; the curtain for the entrance of the Tent; the bronze altar with its bronze grating, its poles, and all its equipment; the wash-basin and its base; the curtains for the enclosure and its posts and bases; the curtain for the entrance of the enclosure and its ropes; the pegs for the Tent; all the equipment to be used in the Tent; and the magnificent garments the priest were to wear in the Holy Place-the sacred clothes Aaron the priest and for his sons. The Israelites had done all the work just as the Lord had commanded Moses.

Moses examined everything and saw that they had made it all just as the Lord had commanded. So Moses blessed them.

Then the last verse *(38) of the book of Exodus says, 'During all their wandering, they could see the cloud of the Lord's presence over the Tent during the day and a fire burning above it during the night.'*

Good fiction! John Grisham and Dan Brown together are no match. God's spokesman and God's designed place of worship, methods, and rules of pleasing the Jewish God are all clearly formalized for the third book 'Leviticus'. It is necessary to state here that at this point of our diagnoses, *the cosmic God* is not yet found in the past two books of Genesis and Exodus. It is hoped, however, that the book of Leviticus will show without doubt God's reality as opposed to the fiction so far exhibited.

Can any reader really accept the above quotations as God's directives to Moses as it is claimed? Or should we not say that every detailed aspect of the instructions is of Moses's own making? That Moses is merely using God's name, as indeed a normal religious practice in every theology, to achieve personal goals? Judge it for yourself!

iii.) Book of Leviticus:

The author of this third book of the Holy Bible is mainly concerned with regulations for worship and ancient religious ceremonies in Israel. This includes duties of rabbis, most appropriate ways for ordinary Jewish people to worship the Holy God of Israel, plus their expected commitment and relationship with him. The Exodus' author had prepared a fabulous stage in the interest of continuity of the original god saga.

Since our main intent is to explore the genuineness of the biblical god, let us see whether this book gives a convincing case for his existence. There are twenty-seven chapters in this book. Leviticus *chapters 1* to 7 deal with laws about offerings and sacrifices. For example, according to Leviticus *1: 1-9,*

> The Lord called to Moses from the 'Tent of the Lord's presence' and gave him the following rules for the Israelites to observe when they offer their sacrifices. When anyone offers an animal sacrifice, it may be one of his cattle or one of his sheep or goat. If he is offering one of his cattle as a burnt-offering, he must bring a bull without any defect. He must present it at the entrance of the Lord's presence so that the Lord may accept him. The man shall put his hand on its head and it will be accepted as a sacrifice to take away his sins. He shall kill the bull there and

the Aaronite priests shall present the blood to the Lord and then throw it against all four sides of the altar which is at the entrance of the Tent. Then he shall skin the animal and cut it up, and the priest shall arrange firewood on the altar and light it. They shall put on the fire the pieces of the animal including the head and the fat.

The man must wash the internal organs and the hind legs, and the officiating priest will burn the whole sacrifice on the altar. The smell of this food-offering is pleasing to the Lord.

Readers of the Bible are required to have faith that this God is the God who created you and me. There are several offerings besides the above; for instance, there is grain-offerings, fellowship-offerings, offerings for unintentional sins, cases requiring sin-offerings, and repayment-offerings. Please, my dear reader, we are told in the Holy Bible that the biblical God is everything; if so, is it really necessary for God to instruct the killing of a goat and to spill its blood in a place of worship? What will be the difference between the worship of this God of Israel and the worship of other gods? Absolutely none! This God appears to be evil and blood thirsty than any other God in other cultures.

Leviticus 7:36 says, 'It is a regulation that the people of Israel must obey for all time to come. The author is clearly using the simplest words to narrate these regulations for Israelites alone. If this God has unlimited powers over everything and everyone, why does he need sacrifices in the first place? How can rational people accept the smell of a burning animal's flesh to be pleasing to God? This is essentially a Jewish tradition which the Nicaea Council deliberately incorporated into the Bible to mesmerize religious addicts.

Chapters 8: 1 to 20 deals with the ordination of Aaron and his sons as priests. These chapters and verses are not so relevant to the present discourse, so we move to *Chapter 11: 1 to 7;* 'Animals that may be eaten; the Lord gave to Moses and Aaron the following regulations for the people of Israel. You may eat any land animal that has divided hoofs and that also chew the cud, but you must not eat camels, badgers or rabbits. They must be considered unclean." Verse 9; "You may eat any kind of fish that has fins and scales: but anything living in the water that does not have fins and scales must not be eaten. Such creatures must be considered unclean".

According to verses 13-22, "You must not eat any of the following birds: eagles, owls, hawks, falcons, buzzards, vultures, crows, ostriches, seagulls, storks, herons, pelicans, cormorants, hoopoes, or bats. All winged insects are unclean, except those that hop. You may eat locusts, crickets, or grasshoppers".

If these Lordly prescriptions were meant for Israelites and observed as such, no one would have questions about it because the author explicitly says it is for Israelites; that is their cultural practice. But because these are Bible texts, most readers may consider these to be God's words and would therefore apply them to their daily life and culture, like everything else in the Bible.

Many Christians have taken these texts as unchallenged heavenly truths applicable universally to all human beings regardless of cultural differences; even the Bible explicitly says it is for Jews.

According to *Chapter15: 1to 2; 'The Lord gave Moses and Aaron the following regulations for the people of Israel. When any man has a discharge from his penis, the discharge is unclean.'*

In *verse 16,'" When a man has an emission of semen, he must bathe his whole body, he remains unclean until evening"*.

Verse 19 says, *"When a woman has her monthly period, she remains unclean for seven days"*.

According to verse 24-25, *"If a man has sexual intercourse with her during her period he is contaminated by her impurity and remains unclean for seven days, and any bed on which he lies on is unclean. If a woman has a flow of blood for several days outside her monthly period or if her flow continues beyond her regular period, she remains unclean as long as the flow continues, just as she is during her monthly period."*

This God of Israel is obviously not the same God in the early part of Genesis, the inventor of Adam and Eve. If he was the same one, the following questions would have been put to him: Why did he make the human genitals if they are unclean? Why mothers and fathers, brothers and sisters, and other relations? Couldn't he just form all of us like he made Adam and Eve? Why a woman's monthly periods, penises and vagina, circumcision and clitoris mutilations? If all these were human undesirables, why are these parts of us? Or should we blame ourselves for possessing them? I am sure you will agree with me that the Holy Bible is full of nonsense. Absolute nonsense indeed, if you think deeply about it!

In Chapter 17: 1-3, "*The Lord commanded Moses to give Aaron, and his sons, and all the people of Israel the following regulations. An Israelite who kills a sheep or a goat as an offering anywhere except at the entrance of the Tent of the Lord's presence has broken the Law*".

Verse 11 says, "*The life of every thing is in the blood, and that is why the Lord has commanded that all blood be poured out on the altar to take away people's sins.*"

A blood-thirsty god! Is he different from idol worshipers? It must be noted seriously that these texts or regulations are intended for the people of Israel. It refers to Israelites' God only: God's people and no one else's god. It will be preposterous for any non-Jewish person to embrace these texts as cosmic directives from the mysterious Heavenly divine Father.

The fiction becomes more interesting in the next chapter. Chapter 18, verses 1-4 talk about the forbidden sexual practices. 'The Lord told Moses to say to the people of Israel',

I am the Lord your God. Do not follow the practices of the people of Egypt, where you once lived, or the people of Canaan, where I am now taking you. Obey my laws and do what I command.'

According to verses 6-23,

> The Lord gave the following regulations. Do not have sexual relations with any of your relations. Do not disgrace your father by having intercourse with your mother. You must not disgrace your own mother. Do not disgrace your father by having intercourse with any of his other wives. Do not have intercourse with your sister or stepsister whether or not she was brought up in the same house with you. Do not have intercourse with your granddaughter; do not have intercourse with a half-sister. Do not have intercourse with your aunt whether she is your mother's sister or your father's sister. Do not have intercourse with your uncle's wife; she too is your aunt. Do not have intercourse with your daughter-in-law, or with your brother's wife. Do not have intercourse with the daughter or granddaughter of a woman with whom you have had intercourse; they may be related to you, and that would be incest. Do not take your wife's sister as one of your wives as long as your wife is living.
>
> Do not have intercourse with a woman during her monthly period, because she is ritually unclean.

The author of Leviticus is obviously a Jew. He or she has demonstrated great skills that he or she understands Jewish customary laws. It will be absurd for us to make these Leviticus or priestly practices applicable to all people just because the Bible says so. As already amplified many times in this discourse, there is nothing holy or sacred in the Holy Bible. It is a man-made thing, a book written with clearly defined objectives—to direct our minds to maintain faith in external invisible images like the Jewish God, and at the same time to freeze our minds from focusing on our inner faith or our inbuilt potentials or capabilities. The Book makes us dependent on others other than ourselves. The Pentateuch had to include this basic Hebrew moral code, without which the entire moral education in both the Torah and the Holy

Bible would be incomplete. There is nothing wrong in equating some of these moral codes with your particular cultural moral code, provided spiritual supernatural links could be voided.

Yes, but this has nothing to do with a cosmic God who is alleged to have instructed Moses to deliver a gospel message to all mankind as most religious leaders are inclined to tell their followers. These commentaries should not be taken as anti-Semitic. They are essentially biblical texts commentary.

Evidently, every culture has its own unique customary laws that regulate sexual practices within its accepted territory. There may be similarities or marginal cultural differences among global cultures. But it would be unethical for any single ethnic group to impose its moral code onto others under the pretext of divine authority as the Jewish book (the Holy Bible) attempts to enforce.

For instance, legal or customary definition of a relative would differ among cultures and countries. Marriage between cousins, nephews, twin sisters with one husband, and many other forbidden biblical Leviticus laws are legitimate customary practices in several cultures around the world. Even among some Hasidic Jews today the ancient biblical sexual code is ignored.

Again it must be emphasized that the author of this book is narrating the imaginary God of Israel's directives to the people of Israel through the main actor—Moses. There is nothing relating to the so-called God of the universe. The author made no such reference. Common sense can tell the reader that the so-called God of Israel is pure fiction, with such acclaimed incredible powers. A real supreme God need not fear any other man-made god such as the God of Israel or any cultural God. So far we have not yet come across any such supreme God. Our search still goes on.

Chapter 18: 21 says, *'Do not hand over any of your children to be used in the worship of the [g]od of Molech, because that would bring disgrace on the name [G]od, the Lord.'* Note that the author used capital 'G' to differentiate the two gods. If this Almighty creator of our universe is a reality, 'fear' as a human behavior cannot be part of his vocabulary, absolutely not. God is supposed to be fearless and invincible. What kind of rubbish is this?

Chapter 19: 4 says, *'Do not abandon me and worship idols; do not make gods of metal and worship them.'*

Worship gods of metal? What is the Cross, the famous symbol of the Catholic Church, statutes of biblical characters such as Jesus Christ, Paul, Peter, Mary, and the human skull image of Mary Magdalene and hundred others in modern Christian Churches?

This God of Israel obviously anticipates that his beloved Israelites can abandon him, make gods of metals, and worship idols. Is the above quotation not sufficient evidence of his holiness's impotence? A God who invents heaven and earth, human beings, and all organic matters on planet Earth, plus a God who is in charge of all actions can alter the above statements and get away with it. How anyone takes this god as his maker or a maker of anything is baffling. Chapter 20: 1-15 gives penalties for disobedience:

> The Lord told Moses to say to the people of Israel, Any of you or any foreigner living among you who gives any of his children to be used in the worship of the god Molech shall be stoned to death by the whole community. If anyone gives one of his children to Molech and makes my sacred Tent unclean and disgraces my holy name, I will turn against him and will no longer consider him one of my people. But if the community ignores what he has done, and does not put him to death, I myself will turn against the man and his whole family and against all who join him in being unfaithful to me and worshiping Molech. I will no longer consider them my people. If anyone goes for advice to people who consult the spirit of the dead, I will

turn against him and will no longer consider him one of my people.

Keep yourselves holy, because I am the Lord your God. Obey my laws, because I am the Lord and I make you holy.

The Lord gave the following regulations: anyone who curses his father or his mother shall be put to death; he is responsible for his own death. If a man commits adultery with the wife of a fellow Israelite, both he and the woman shall be put to death. A man who has intercourse with one of his father's wives disgraces his father, and both he and the woman shall be put to death. If a man has sexual relations with another man, they have done a disgusting thing, and both shall be put to death. If a man has sexual intercourse with his daughter-in-law, they shall both be put to death. They have committed incest and are responsible for their own death. If a man marries a woman and her mother all three shall be burnt to death. If a woman tries to have sexual relations with an animal, she and the animal shall be put to death. If a man has sexual relations with an animal he and the animal shall be put to death.

Verse 27 ends this chapter with *'Any man or woman who consults the spirit of the dead shall be stoned to death.'*

Well, these are some of the words the God of Israel tells his people (Israelites) as according to the author of the book of Leviticus. Again it must be noted that no single reference is made to a *cosmic god*. This Jewish god instructed Moses, who, we are already informed by the book of Exodus, was a Jewish leader to his people. All the penalties for those who break his holiness's laws are stated herewith.

It becomes very confusing when we compare the powers of the god in Genesis and the god in Leviticus. The God in

Genesis *appears* to possess supreme powers including making and unmaking humans with ordinary soil and a single human rib. He must have made the devil himself too, even though the writer did not mention it per se, but how can the god in Leviticus be worried if an Israelite chooses to worship the spirit of a dead person or an idol such as Molech? What happened to his godly powers? He could have made them to evaporate instantly. Should God really be worried about man's disobedience? I don't get it. What makes god a God is his overall cosmic superiority; his supreme Lordship is supposed to conquer all without exception. Leviticus' God is a coward. He is certainly frightful of his own creation. Was he not aware of what he was creating, this heterogeneous intelligent individual in possession with organic brain, a real rational human being?

Seriously, so far, from Genesis to Leviticus, there is not a single convincing case to support the validity of a unique super God, a maker or a creator of anything or anyone that is not man-made. The author concludes with God of Israel's commands or laws given to Moses on Mount Sinai for the people of Israel. These include complete jurisprudence on crimes and civil cases plus God's stipulated punishments such as death penalty, fines, banishment, retribution, and community trials. God of Israel acts as the invisible chief director for Israeli affairs, with Moses acting as the visible chief executive, the legislator, the judge, and simultaneously, the spiritual agent linking God and his chosen people of Israel.

The main actor here is Mr. Moses. He does not pretend to be a God. He acts as a faithful messenger of the Jews' God, indeed not a cosmic God by any definition. By this narrative, Mr. Moses undoubtedly sees God as no one short of Israelites' God. It will not only be a misunderstanding of these ancient scriptures to make him the universal God but rather a wish to brand this God as the universal God. For those of you, who are undoubtedly hungry for a single universal God's existence, please search somewhere else for that entity, plausibly your localized cosmic or cultural God.

As already strongly argued, every human community creates and worships its own cultural God, just like the God of Israel, which was created by Jewish ancestors such as Abraham and Jacob's generation for Jews. At the end of your search, I can bet you that you will end up with either your cultural God or succumb to someone's cultural God.

As written in *Isaiah 29*, and especially *verse 13*, their religion is nothing but human rules and traditions. These precepts, which are popularly termed as God's own words, such as the Ten Commandments, were formulated by Moses and the Jewish elders to perpetuate ignorance and to conceal the truth which would have otherwise revealed the fruit of the truth within us. In the Gnostic Bible and according to the Gospel of Mary Magdalene, *'Jesus was asked how a person sees a vision; through the soul or the spirit? The savior answered saying; a person sees neither through the soul nor the spirit. The mind which lives between the two sees the vision. Where the mind is, there is the treasure'*.

The bottom line of everything is the human brain, the organ which captures information and has it processed *into* mind.

iv.) Book of Numbers:

The book contains thirty-six chapters. The author tells the story of the Jews' forty years trip from Mount Sinai to the eastern border of the land their God had promised to give them. The curious name of the book refers to an important feature of the story. Moses counted all the Israelites before departure from Mount Sinai and did another census when they arrived at Moad, east of Jordan about a generation later. The author narrates difficulties which the Jews encountered, how they became discouraged and rebellious against their God and their leader Moses. The book also talks about God's faithfulness and persistent care of his people in spite of their weakness and disobedience, and Moses's steadfastness and often impatience and devotion to both God and his people.

Numbers Chapter 1:1-5 says, *'On the first day of the second month in the second year after the people of Israel left Egypt, the Lord spoke to Moses there in the Tent of his presence in the Sinai desert. He said, 'You and Aaron are to take a census of the people of Israel by clans and families. List the names of all the men twenty years old and older who are fit for military service. Ask one clan chief from each tribe to help you.'* Verses 53 says, *But the Levites camp round the Tent to guide it, so that no one may come near and cause my anger to strike the community of Israel'.*

In chapter 3; 10, *'You shall appoint Aaron and his sons to carry out the duties of the priesthood: anyone else who tries to do so shall be put to death.'*

According to Numders 12: 1-10,

> Moses had married a Cushite woman (a non-Jew) and Miriam and Aaron criticized him for it. They said, Had the Lord spoken only through Moses? Hasn't he also spoken through us? The Lord heard what they said. (Moses was a humble man, more humble than any one else on earth). Suddenly the Lord said to Moses, Aaron and Miriam, 'I want the three of you to come out to the Tent of my presence.' They went and the Lord came in a pillar of cloud, stood at the entrance of the Tent and called out, Aaron! Miriam! The two of them stepped forward, and the Lord, Now hear what I have to say. When there are prophets among you, I reveal myself to them in visions and speak to them in dreams. It is different when I speak with my servant Moses; I have put them in charge of my people Israel. So I speak to him face to face, clearly and not in riddles; he has even seen my form! How dare you speak against my servant Moses? The Lord was angry with them and so as he departed and the cloud left the Tent, Miriam's skin was suddenly covered with a dreaded disease and turned as white as snow.

According to *Exodus 2: 10*, the name Moses means 'pull out,' in Egyptian language. It was given to him by the Egyptian princes who found and picked him up at the river bank. It is an Egyptian name, and there is no certainty that Moses was a Jew. Again, by the above quote, Aaron and Miriam are offended by his marriage to a non-Jew. Once again, God of the Jews is being ungodly and emotional.

According to Numbers 14; 26 to 29:

> The Lord said to Moses and Aaron, How much longer are these wicked people going to complain against me? I have heard enough of these complain! Now give them this answer: I swear that as surely as I live, I will do to you just what you have asked. I the Lord have spoken. You will die and your corpses will be scattered across this wilderness. Because you have complained against me, none of you over twenty years of age will ever enter that land.

According to verses 34 and 35, *'You will know what it means to have me against you! I swear that I will do this to you wicked people who have gathered together against me. Here in the wilderness every one of you will die. I the Lord have spoken.'*

Even Moses was surprised of the ungodly behavior of the god of Israel when he asked, in chapter 16: 22-23, " .... *O God you are the source of all life. When one man sins, do you get angry with the whole community?'* A swearing god, an angry god, a complainant god, and a revengeful god of Israel is not fit by any definition to be the God for mankind. Even Moses with human brain is questioning God's prudence.

The author of the book of *Numbers 23; 19* says *'God is not like men who lie. He is not a human who changes his mind.'*

But the same author in chapter 25: 10 says,' .... *I am no longer angry with the people of Israel.'* The above quotations are definitely contradictory.

The author is obviously putting words into the so-called God's mouth. Or perhaps he or she is confirming our long time suspicion that the so-called God is unreal and phony. This is without question kids' stuff and very repulsive. God becomes a military commander, orders Moses to attack the Midianites, and destroy them;

Numbers 31: 15 says, *'Why have you kept all the women alive?'* Numbers 31: 17-18 says, *'Kill every boy and kill every woman who has had sexual intercourse. But keep alive for yourselves all the girls and all the women who are virgins'*. How on earth can this character be the maker of all human beings?

In Chapter 32, the Jewish God got furious on several occasions with death threats on anyone who disobeyed his Lordship's commandments. The God of Israel gave to the tribes of Reuben and Gad the towns of Ataroth, Dibon, Jazer, Nimrah, Heshbon, Elealeh, Nebo, and Beon; but these towns were already occupied by non-Jewish indigenous tribes. The Lord again gave the following instructions to Moses for the people of Israel.

'When you cross the Jordan into the land of Canaan you must drive out all the inhabitant of the land. Destroy all their stone and metal idols and all their places of worship. Occupy the land and settle in it, because I am giving it to you' (Num 33: 51-53). Are those non-Jews also not part of the children of the beloved God, the almighty creator? According to Numbers 34: 16-29, *'The Lord said to Moses, Eleazar the priest, and Joshua son of Nun will divide the land for the people. Take the leader from each tribe to help them divide it. These are the men the Lord chose:'*

Tribe	**Leader**
Judah	*Caleb son of Jephunneh*
Simon	*Shelumiel son of Ammihud*
Benjamin	*Elidad son of Chislon*
Dan	*Bukki son of Jogli*
Manasseh	*Hanniel son of Ephold*

Ephraim	*Kemuel* son of Shiphtan
Zebulon	*Elizaphan* son of Parnach
Issachar	*Paltiel* son of Azzan
Asher	*Ahihud* son of Shelomi
Naphtali	*Pedahel* son of Ammihud.

These are the men that the Lord assigned to divide the property for the people of Israel in the land of Canaan'.

The amazing similarity of the book of *Numbers* and the book called *Mein Kampf* written by Adolf Hitler in 1925 when he was in prison is unbelievable. Readers who are familiar with these two books will notice that Hitler's *Lebensraum* ideas although originated in the mid-nineteenth century by Friedrich Razel, are almost identical with the God of Israel's land policies in favour of the people of Israel. It was Hitler who improved and implemented the land acquisition ideas proposed by Friedrich Razel in Europe during his political leadership in Germany between January 1933 and April 1945.

'Lebensraum' is a German word for 'habitat' or a 'living space'. It states that 'development of a people was primarily dependent on their geographical situation and that a people that successfully adapt to one location would proceed naturally to another to prosper. Expansion into Eastern Europe, Scandinavian countries, Alsace, and Lorraine of France, Belgium, Poland, and the then Soviet Union was a natural necessity of German people to acquire'. It was stated as a 'Nazi party policy to acquire by any means possible, *i.e.* to kill, to deport, to enslave, or to exterminate by starvation the lands occupied by inferior people to the benefit of Germanic superior race'. Just as the God of Israel saw the land belonging to Moabites, Hittites, Edom, Zin, Kadesh, Amorite, Og, Mount Pisgah, Mount Sinai, and Canaan, as a natural necessity for the Israelites. For instance, in Numbers 32: 3 to 4, *'This region which the Lord has given to Israelites to occupy, the towns of Ataroth, Dibon, Jazer, Nimrah, Heshbon, Elealeh, Sibmah, Nebo and Beon is good land for cattle, and we have many cattle:'*

Verse 7 says, *'How can you discourage the people of Israel from crossing the Jordan into the land the Lord has given them.'*

These lands are owned and occupied by the local people, but God of Israel in his own might has ordered the slaughter and complete extermination of innocent inhabitants for the benefit of his chosen people. Just as Hitler did between 1933 and 1945 as mentioned above simply because he, Adolf Hitler, the number one twentieth century monster, or the God of Israel as portrayed in the book of Numbers, sees it as a natural necessity to exterminate millions from their lands for the benefit of his so-called Germanic or Jewish superior race.

In *Mein Kampf*, Hitler wrote, 'Without consideration of traditions and prejudices, Germany must find the courage to gather our people and their strength for an advance along the road that will lead this people from its present restricted land space to new land and soil and hence also free it from the danger of vanishing from the earth or of serving others as a slave nation.'

The God in the book of Numbers can never be described as a God for all people. Even though he described himself as God for the Jews and certainly behaved as a selfish god only for a select few, the Jews, Christians all over the world have taken him as the god for everyone or as the God described in the first eleven chapters of Genesis—the creator. These are two distinct characters. Neither can be transformed into the other nor can either be considered as the designer of the universe. The Jewish God in the book of Numbers is demonstrably a racist brute, a character that is even more vicious than Adolf Hitler. It will be preposterous for any Bible reader to compare a creator god with a destroyer god, or indeed the conclusion of voiding all ideas of god's existence becomes self-evident. It is unfortunate though that biblical writers draw no distinction between this infamous God of Israel and the God they are trying to convince you and me as our maker.

There is certainly not the slightest evidence in the Holy Bible that suggests the possibility of an entity, imaginary or really worthy of the title 'Cosmic God'; absolutely zero. None! I challenge every reader to disregard my remarks and to reread 'Numbers' and judge it yourself.

v.) Book of Deuteronomy (God of Israel)

This is the fifth book of the Holy Bible. It is also the last book of the five books which were directly copied from the Jewish Bible, 'the Torah', into the Christian Holy Bible by the Council of Nicaea. The remaining thirty-four books of the Old Testament equally form part of the Torah except that they represent parts of the books. Briefly the book contains a chain of lectures given by Moses, the main actor or character, to the Israelites—the God's chosen people in the land of Moab.

For our convenience, the book will be divided into four main parts:

1. Moses's recollection of great events during the past forty years. His appeal to the people to remember how god had led them through the wilderness and to be obedient and loyal to God.
2. Moses reviews the Ten Commandments and emphasizes the meaning of the First Commandment, calling the people to devotion to the Lord alone. The various laws that are to govern Israel's life in the Promised Land are reviewed.
3. Moses reminds Israelites the meaning of God's covenant with them and calls for them to renew their commitment to its obligations.
4. Lastly, Joshua is commissioned as the next leader of God's people. After singing a song celebrating God's faithfulness and pronouncing a blessing on the tribes of Israel, Moses dies in Moab, east of river Jordan. He was buried by God himself in the mountains at age 120 years as according to Deuteronomy 34, 5-7.

A major theme of Deuteronomy is that God has saved and blessed his chosen people whom he loves; in return his people are to remember, to love, and to obey him so that they may have life and continued blessings. The key verses of the book are:

Deuteronomy 6: 4-6: *'Love the Lord your God with all your heart, with all your Soul, and with all your strength.'* This of course is Jesus' favorite commandment as exhibited in several books of the New Testament.

Our main quest at this time is to identify the existence of our supposed maker or the cosmic God as according to the author of the book of Deuteronomy.

Deuteronomy 1: 6-8:
> When we were at Mount Sinai, the Lord our God said to us, 'you have stayed long enough at this mountain. Break camp and move on. Go to the hill country of the Amorites and to all the surrounding regions—to the Jordan valley, to the hill—country, and the low lands, to the southern regions, and to the Mediterranean coast. Go to the land of Canaan and on beyond the Lebanon Mountains as far as to the great River Euphrates. All of this is the land which I, the Lord, promised to give to your ancestors, Abraham, Isaac, and Jacob, and all their descendants. Go and occupy it.'
>
> Clearly this is the voice of God of the Jews; the selfish fuehrer himself, who is solely concerned with the Jews. He cares least of the people who already occupied the lands stated above. He gave his self-acclaimed Almighty command *'Go and occupy it'*. And the world refused to question who the hell this god thinks he is. How on earth can any practical person conclude that a God for all mankind can exterminate any ethnic humans he has created in their natural habitat in favor of another group without a reasonable course? If this

God created all humans, why does he love only Jews as his chosen people and deliberately destroys others for inexplicable reason?

Deuteronomy 2: 24-36:

After we had passed through Moab, the Lord said to us, Start out and cross the River Arnon. I am placing in your power Sihon, the Amorite King of Heshbon, along with his land. Attack him and begin occupying his land. From today on I will make people everywhere afraid of you. Everyone will tremble with fear at the mention of your name. then I will send messengers from the desert of Kedemoth to King Sihon of Heshbon with the following offer of peace: 'let us pass through your country.

All we want to do is to pass through your country.

But Sihon would not let us pass through his country. The Lord your God had made him stubborn and rebellious, so that we could defeat him and take his territory, which we still occupy.

Then the Lord said to me, 'Look I have made King Sihon and his land helpless before you; take his land and occupy it. Sihon came out with all his men to fight us, near the town of Jahaz, but the Lord our God put him in our power and we killed him, his sons, and all his men'. At the same time we captured and destroyed every town, and put everyone to death, men, women, and children. We left no survivors. We took the cattle and plundered the towns. The Lord our God let us capture all the towns from Aroer, on the edge of the valley of Arnon, and city in the middle of that valley, all the way to Gilead. No town had walls too strong for us.

Deuteronomy 3: 1-7:
>Next we moved north towards the region of Bashan, and King Og came out with all his men to fight us near the town of Edrei. But the Lord said to me, 'Don't be afraid of him. I am going to give him, his men, and all his territory to you. Do the same to him as you did to Sihon the Amorite king who ruled in Heshbon. So Lord also placed King Og and his people in our power, and we slaughtered them all. At the same time that we captured his towns—there was not one that we did not take. In all we captured sixty towns. We destroyed all the towns and put to death all the men, women, and children, just as we did in the towns that belonged to King Sihon of Heshbon.

Deuteronomy 4: 15-19:
>When the Lord spoke to you from the fire on Mount Sinai, you did not see any form. For your own good then make certain that you do not sin by making for yourselves an idol in any form at all—whether man or woman, animal or bird, reptile or fish. Do not be tempted to worship and serve what you see in the sky—the sun, the moon, and the stars. The Lord your God has given these to all other peoples for them to worship.

Deuteronomy 4: 24: *'Because the Lord your God is like a flaming fire, he tolerates no rivals.'*

Deuteronomy 5: 7: *'Worship no god but me.'*

Deuteronomy 7: 6: *'Do this because you belong to the Lord your God. From all the peoples on earth he chose you to be his own special people.'*

Deuteronomy 14: 25-26: *'Sell your produce and take the money with you to the one place of worship. Spend it on whatever you want beef, lamb, wine, beer—and there in the presence of the Lord your God, you and your families are to eat and enjoy yourselves.'*

In several chapters and verses, such as Deuteronomy chapters 19 through to chapter 34, the God of Israel endorses slave trading, merciless torture and killings and several atrocities. Punishments such as, *'life for life, eye for eye, tooth for tooth, hand for hand, and foot for foot,* in Deuteronomy are grossly ungodly. In *Chapters 19, 20, and 21,* this bloodthirsty God of Israel's behavior, as narrated by the author of the book of Deuteronomy, totally ignores the basic theological and moral tenet of an 'Omni' God. This very god has laid down his own ten commandments in Exodus 20 plus additional 600 other laws for the people of Israel to obey, yet he, God, is the first one who breaks all of them. If indeed there will ever be a judgment day, the Lord God of Israel must be the first to be brought to justice for the loss of multitudes of innocent life that took place under his command. The following statement is surely true, *'The people of Israel, no god is like your God' (Deut 33: 26),* Indeed, *'there has never been a prophet in Israel like Moses; the Lord spoke with him face to face' (Deut. 34: 10),* and made him the biggest monster ever narrated in any fiction book. This character was responsible for the massacre of thousands of peaceful innocent people in several cities and towns along their route to the so-called Promised Land of Canaan.

I sincerely hope that every reader will henceforth spare some moments after reading this discourse to go back to re-read the book of Deuteronomy for personal analysis and judgment, if my demeaning conclusion is unjustified. The entire text is grossly sadistic and unholy. It is quite surprising that Constantine and his councils of bishops should include this portion in a book which is intended to be worshiped as a holy book, perfect in goodness and righteousness. There is nothing good or righteous about this sheer wickedness, no matter who commits it.

vi.) Book of Joshua (God of Israel)

This sixth book of the Old Testament also shows up in the Jewish Bible, 'The Torah'. It is the first of the *oral* texts as opposed to *written* texts of the Torah. Joshua was the first of the Jewish prophets beside Moses who was appointed by the Jewish God to lead the Israelites' invasion and occupation of Canaan.

He was Moses's successor and God's commander-in-chief destined to annihilate all the inhabitants of the lands of east and west of Jordan, the cities of the Levites, Hittites, Acacia, Jericho, and many more. This book is obviously a continuation of the Jewish God's genocide and land acquisition policies in favor of his beloved Israelites. Can this God of Israel really be the God of all peoples of the world? How anybody can justify these barbaric evil deeds, sanctioned by a benevolent cosmic God, is incomprehensible.

In Joshua 3: 7-10:

> The Lord said to Joshua, What I do today will make all the people of Israel begin to honour you as a great man and they will realize that I am with you as I was with Moses. Then Joshua said to the people, come here to listen to what the Lord your God has to say. As you advance, he will surely drive out the Canaanites, the Hittites, the Perizites, Girgashites, the Amorites, and the Jebusites. You will know that the living God is among you.

Joshua 4:13:
In presence of the Lord, about forty thousand men ready for war crossed over to the plain near Jericho.

Joshua 5; 2-3: The Lord said to Joshua, make some knives out of flint and circumcise the Israelites. So Joshua did as Lord had commanded, and he circumcised the Israelites at a place call Circumcision Hill. Verse 6: *for during the forty years the people spent crossing the desert, the baby boys had not been circumcised.* Verse 8-9: *After the circumcision was completed the whole nation stayed in the camp until the wounds had healed. The Lord said to Joshua, today I have removed from you the disgrace of being slaves in Egypt.*

Chapter 6; 17: *the city and everything in it must be totally destroyed as an offering to the Lord.*

Verse 19: *Everything made of silver, gold, bronze, or iron is set apart for the Lords' treasury.* Verse 21: *With their swords they killed everyone in the city, men and women, young and old. They also killed the cattle, sheep and donkeys.* Verse 26: *At that time Joshua issued a solemn warning: 'Anyone who tries to rebuild the city of Jericho will be under the Lords curse'.*

Circumcised people or Jews are considered honored and liberated in the eyes of God of Israel. The author of the book of Joshua, whoever that might be, is telling us both Christians and non-Christians that the supposed creator of the world, the Israeli God, deliberately designed male's penis as an impure or disgraceful organ which requires circumcision before the person is recognized by God. Indeed, the use of the generic word 'World' in the Holy Bible must be understood to represent the Hebrew world, 'Israel' only. It certainly would make sense if the entire Bible, not just the book of Joshua, is accepted as a narration of Jewish cultural experiences. These experiences may be similar to other cultural practices, but to take these as the generic human experience is a complete misunderstanding of biblical scriptures. The Holy Bible itself repeatedly mentions this that the supreme divine creature the book talks about is the God of Israel and that 'He' has chosen Jews as His favorite people with a written covenant linking 'Him' and 'Them'.

Now let us examine God's attitude towards the non-Jewish tribes mentioned by this author and repeated here—the Canaanites, the Hittites, the Perizzites, Girgashites, the Amorites, the Jebusites' Ai, Hevitis. Without the slightest hint of provocation narrated by the author, the God of Israel commanded his chosen Israelite leader, Mr. Joshua, the son of Nun, to form an army of thirty thousand *circumcised* strong men (*Joshua 8: 1-29*). The only use of the army is to kill all the indigenous people mentioned above, take their jewelry, burn their houses, and take possession of their land, and also in some instances, torture and humiliate their kings before assassinating them.

If this monstrous inhuman act of a god who is worshiped by nearly one-third of current world population can be rationally defended, how implausible may this same people not find justification for monster Adolf Hitler's genocidal and land policies during the Second World War. Can any one in his or her right senses come forth with a justified case for God? It is not rationally acceptable just to say that he is the giver of all things including life, and can therefore take it at any time for no reason. This will be a naive as well as irrational excuse, because right from Genesis to Joshua, no God other than this same Jewish God is surprisingly claimed by the authors to have invented all things natural. The hypothesis that this same God is the architect of the world is extremely weak, infantile, and complete gibberish. This God has emphatically stated in the Song of Moses in Deuteronomy, chapter 32, and elsewhere that the *God is for Israelites only and not for any other ethnic people.* Christians with non-Jewish background find it much harder to accept this; they find it easier to chew a rock than to absorb this clear observation that the biblical God is not for non-Jews.

In Joshua 10: 10-12, the author continues,

> The Lord made the Amorites panic at the sight of Israel's army. The Israelites slaughtered them at Gibeon and pursued them down the mountain pass at Berth Heron, keeping up the attack as far south as Azekah and Makkedah. While the Amorites were running down the pass from the Israelites army, the Lord made large hailstones fall down on them all the way to Azekah. More were killed by the hailstones than by the Israelites.

Joshua 11: 8-9, *'The fight continued until none of the enemy was left alive. Joshua did to them what the Lord had commanded . . .'*

Joshua's atrocity on innocent multitude, as commanded by God, throughout the remaining chapters of the book of Joshua, is too overwhelming to be copied here. It is strongly recommended that the reader reads the book directly from the so-called sacred Holy Bible for personal assessment. I also like to draw the attention of the reader of this book to some simple questions about the nature of the God portrayed so far by its various anonymous authors. First, in the book of Genesis from chapter one to chapter two, verses twenty-five, the author made some efforts to explain how God, 'a He not a She', from nowhere created the universe out of nothing. Then this miraculous 'He' God invents all organic maters including human beings in the presence of unidentified others.

Beyond Genesis 2: 26, God, the supposed creator and owner of the universe, became invisible and disappeared from the visible world. The authors, however, inform readers that his departure was caused by the sin of a single matter 'Eve', a female human being he, God, himself created out of sympathy. This same incredible God confessed later in the same book of Genesis his regret for creating humans. From that period to date, this author's God has lived in his invisible world to conduct the affairs of the visible world he personally invented. We are informed by the remaining sixty-five biblical authors that this same original primordial God transformed to become God 'only' for Israelites. The racist brute who willfully massacred thousands of innocent non-Jews, and who for no fault of theirs happened to occupy fertile land worthy of Jewish settlement.

We shall also be informed in the second phase of this discourse, the New Testament diagnosis in which this same invisible Jewish God becomes a human being, someone, a Jew, called Jesus. Like a frog-egg into a fish-like creature with a fish tail that swims and which slowly mutates into a four-legged frog, Jesus became the Christ who was born like you and I, from a woman's vagina; and again into Jesus Christ after baptism who became the son of God, and matured into God the creator himself. This is the incredible mythological creature which the Holy Bible entices humanity to swallow without challenge. I find it very hard to figure out how the God in Genesis can transform to an ethnic God several millions of years later and again transforms into a real human being (Jesus) after several thousands of years; thirty-three years during Christ Era, this man-made visible God (Jesus) dies and resurrected into the original visible Genesis God (Christ): at same time with the image of the brute ethnic God of Israel. How can we not believe these transformations to be fabricated stories?

Several historical and archeological evidence exist which supports the understanding of biblical godly stories as fabricated. The recorded history of Holy Bible itself indicates that none of the main characters in the book wrote any document of any kind before they died. And I would assess that over 98 per cent of all biblical stories are secondhand information written by superstitious, enthusiasts, the clergy, and interest groups whose livelihood have depended on the biblical stories per se.

vii.) Book of Judges (God of Israel)

This is the seventh book of the Old Testament. The author's name is not mentioned anywhere. The book's title is 'Judges'. The name is misleading in this context. The book's title has nothing to do with judges in the legal sense of the word. This book is mainly *about Jewish heroes*, essentially military or charismatic leaders, who are characterized as judges. The book tells stories of Israelites' invasion of Canaan and their establishment of a monarchy. The best known among them was Samson, whose works are recorded in chapters 13 to 16. The book narrates that Israel's survival was made possible because of Jews' loyalty to the *God of Israel*. Although their disloyalty always led to disaster, the God of Israel was always ready to save his chosen people, who turned to him for help.

The main theme of this work is the essence of God. It is hoped, however, that by reading the Holy Bible a clear picture of biblical God would be exposed and thereafter share the idea with readers. But one thing that has become more apparent so far as up to the seventh book (Judges), is the persistent use of *ambiguous Hebrew words and terminologies* by biblical authors to narrate Bible stories. Again it is noted that the choice of words used then and date of events narrated by these authors, especially the Old Testament, reflecting Iron Age period, about 1,200BC, are different from the words used and event dates AD1000. It is even far more different and difficult in the twenty-first century when trying to understand those prehistoric words in the light of contemporary discourses.

For instance, such Hebrew words as 'Israel' then meant 'he struggles with God' or 'God struggles.' It did not represent a name of a country with geographically defined boundary clearly marked on the globe, and with well-structured governmental institutions as it is today. Israel probably represented a tribe or an ethnic group of people with common cultural heritage, language, and traditions.

In today's understanding, the word 'judge' implies a court official who is appointed to sit as a presiding judge or officer in charge of hearing disputes between plaintiff and accused and deciding who is guilty or innocent in a court of law.

A judge in the Iron Age period was a self-appointed or a Yahweh tribal selected leader of a Jewish community. He was accountable only to God of Israel. Hebrew word for a judge then was (*sh-p-t*). The real and the only judge, then, was God. Yahweh became *shophet* in Hebrew in the later years. Similarly, such words as king, messiah, and rabbi, as used in the Holy Bible had multiple meanings then, compared to their modern definitions. Historically, Israel never had a king, not even during their alleged forceful occupation of the promised land of Canaan by modern definition. A king is supposed to be a male monarch who governs a defined geographical territory. His position must be hereditary and destined to rule for life. By this modern understanding none of the biblical kings of Israel qualifies for a king's title. Thus the names given in the book of Judges as kings represent self-styled war leaders, or charismatic gallant men, who professed to be Yahweh's beloved appointed leaders. All their successes and failures were Yahweh's commandments.

The first Judge mentioned in the book of Judges 2. 6-10 is Othniel; a nephew of Caleb became a charismatic hero after a successful battle. He is ranked as next to Joshua in warfare. Among Israeli tribes, Ehud is mentioned in Judges 3. 12-30 as remembered for his craftiness in killing the enemy Moabites' king, and for his left-handedness. Deborah, a female heroine, is also a classified judge, probably because of her prophetic powers.

Several judges or heroes mentioned in this book are no different from the previous Jewish leaders. They are all God of Israel appointed assassins. Their names appear in the Holy Bible simply because they were brilliant killers appointed by the God of Israel, who we are told is the same universal God who created our ancestors billions of years ago, and perpetually continues to create human beings including you and I.

Samson is projected in this book as one-of-a-kind Jew for his impeccable military might in *Judges 13. 5-16*. But as far as this inquiry shows, his ungodly brutal acts deserve no praises. He will be among the first wicked lot to be tried if there is such a thing as a universal judgment day.

In the conclusion of this book in our quest for a cosmic creator, the author failed to show anything, historically, geographically, philosophically, and theologically to support a case for the existence of an entity which is capable of inventing anything natural.

One thing that is quite apparent in the book of Judges is all the twelve tribes of Israel appeared to be disobedient to their God, when they were instructed by God to kill every inhabitant of their newly acquired territories. Some tribes decided not to do so. God decided to punish his own beloved people by making them fight among themselves; Jews against Jews. Judges chapters 18-21 tell stories of squabbles and heinous atrocities amongst Israelites, all under the commandments of their own God who we are told is the same God who created mankind. For instance, Judges 20: 18 says, *'The Israelites went to the place of worship at Bethel, and there they asked God, which tribe should attack the Benjaminites first? The Lord answered the tribe of Judah.'* Benjamin's army killed 22,000 Israelites. *Verses 23 and 24 narrate the following:, 'They asked God, should we go again into battle against our brothers the Benjaminites? The Lord answered yes'.* This time 18,000 well-trained Israelite soldiers were killed by the tribe of Benjamin's troops.

Verse 28: *'The people of Israel asked the Lord the third time, should we go out to fight our brothers the Benjaminites again or should we give up? The Lord answered, fight. Tomorrow I will give you victory over them'.*

Verse 32: *'The Benjaminites said, we have beaten them just as we did before.'* Israelites were again defeated; this time only thirty Israelites were killed. Then finally, verse 35 says, *'The Lord gave Israel victory over the Army of Benjamin. The Israelites killed 25,100 of the enemy that day, and the Benjaminites realized they were defeated.'*

Now let's find out what the author of the book of Judges is saying in the last chapter, Chapter 21: 1-12,

> When the Israelites had gathered at Mizpah, they had made a solemn promise to the Lord: none of us will allow Benjaminites to marry a daughter of ours.' So the people of Israel went to Bethel and sat there in the presence of God until evening. Loudly and bitterly they mourned: Lord God of Israel, why has this happened? Why is the tribe of Benjamin about to disappear from Israel? Early the next morning, the people built an altar there, offered fellowship sacrifices and burnt some sacrifices whole. They asked, is there any group out of all the tribes of Israel that did not go to the gathering in the Lord's presence at Mizpah? (They had taken a solemn oath that anyone who had not gone to Mizpah would be put to death.) The people of Israel felt sorry for their brothers the

Benjaminites and said; today we shall go to provide wives for the men of Benjamin who are left. We have made a solemn promise to the Lord that we will not give them any of our daughters. When they asked if there was some group out of the tribes of Israel that had not gone to the gathering at Mizpah, they found out that no one from Jabesh in Gilead had been there; at the roll call of the Army no one from Jabesh had responded. So the assembly sent 12,000 of their bravest men with orders 'Go and kill everyone in Jabesh including women and children. Kill all the males and also every woman who is not a virgin.' Among the people in Jabesh they found 400 young virgins, so they brought them to the camp at Shiloh, which is in the land of Canaan.

Verses 20, 21, 22: *'They said to the Benjaminites . . . Each of you take a wife by force from among the girls and take her back to the territory of Benjamin with you.'* The last verse (verse 25) of the book of Judges says, *'There was no king in Israel at the time. Everyone did just as he pleased.'*

I am wondering what our Christian readers think about the entire book of Judges, particularly the last chapter quoted above. It would have been imagined to be a summary of Jewish ancient cultural history as presented by anonymous ancient authors in a holy book. A critical examination of the contents reveals a lot of puzzles. But besides the brainteasers, one thing is absolutely certain. That is, the universal God or the Omni, Alpha and Omega, which past and present generations of humanity have been indoctrinated as the designer of 'all' human beings and 'all' things natural, is completely absent from this book. There is no trace of that God in Judges or any of the books anywhere.

The author made no effort at all to promote the God of Israel or any other God as compared to the preceding six books authors in the Old Testament already reviewed.

The first riddle of this book is the placement of the book. Why is it the seventh book and not the sixth or the eighth book, or indeed anywhere among the thirty-nine old text books? The second puzzle is the book title, 'Judges'; why not heroes or thugs? It seems to me that the editors of these texts deliberately searched for words with multiple meanings, purposely to provide alternative or multiple interpretations of biblical texts.

Holy Bible readers are informed that none apart from God is capable of judgment. The heavenly God is the ultimate *shophet*. He is the only *Judge*, or the supreme arbiter who judges what is good and what is bad, what is right and what is wrong. This narrative is prevalent in both Old and New Testaments; hence there can be no denial that the author of this book was unaware of the multiple meanings of the book title 'Judges'. It was a calculated choice, possibly engineered by Irenaeus, the second century bishop of Lyons to enhance future multiple interpretations and adaptations of these texts.

The central theme as previously stated is deeds of Jewish heroes of the period, and there is a Hebrew word for heroes. A choice to avoid the use of the appropriate word only indicates the existence of an ulterior motive. Irenaeus wrote several theological documents which stated that it is not the business of the church to concern itself with heroes. It is rather the works of heroes that counts. And for this reason, comprehensive written biographies of these Jewish icons or personalities are not available. Biblical stories are told as tales to fulfill the churches' agenda. By adding letter 's' to the word 'Judge', for example, sets a deliberate platform for confusion and uncertainty to keep readers' mind out of focus. Who are the judge(*s*)? Are they the new generation of Israelites themselves, or the Moabites' Gods whom the God's people at that time chose to worship, instead of worshiping the God of Israel?

Now, let us attempt to find out the reason why the book of Judges needs to be placed as the seventh but not the sixth or the eighth book of the Old Testament. The seventh book is essentially a summary of Godly images and their works in accordance with the Jewish culture. Note that *Genesis* deals with the supposed creation of the world and the beginning of life as understood by ancient Jews. It concludes with works of the transformed God (the creator) into the God of Israel who functioned as the commander-in-chief of the nation of Israel.

Exodus continues with problems Israelites encountered in transit to their promised land under the guardianship of their God. *Leviticus* projects the God of Israel's wickedness and senseless act of brutalities just to pave way for his chosen people. *Numbers* continues with the same brute commandments from God to guide Moses towards Mount Sinai after taking census of the new generation of Israelites. Both God and Jews show to each other their worst characters—God's impatience and Israelites disobedience—before crossing River Jordan.

Deuteronomy demonstrates the determination and will of the Israelites to reach the Promised Land at all costs. Moses, the obedient servant of the God of Israel, after almost forty years in the wilderness performed his last but worse satanic duties in the land of Moab. He obtained the Ten Commandments as delivered by the God of Israel. He gave the last pieces of lectures to his people and died in Moab at the mountain top at the age of 120 after commissioning Joshua as his successor, according to the will of God of Israel. Moses had taught his people all they needed to know as God's children of Israel. Book of *Joshua* narrates the last of the Jewish holy wars, the conquering and occupation of Canaan by God's beloved children of Israel.

After all the trials and tribulation *from Genesis 2: 27 to Joshua 24,* a summary of Jewish heroes who made these possible is needed as a book. The title 'Judges' instead of heroes therefore represents, one, the total events of the transition of Jewish people and their relationship with God of Israel, and two, the last transformational bridge for God of Israel becoming the Universal 'human' God (Jesus Christ) as exposed in the New Testament. The Genesis' God, God of Israel, and Jesus Christ as the latter-day God must necessarily be unified. We shall see if they succeed.

CHAPTER 7

God According to New Testament

i.) The Book of Matthew (Son of God)

The thirty-nine books of the Old Testament have been reviewed with the single aim of searching for a universal single God, a God who could be claimed to be the maker of all living organisms including human beings. The outcome of the search shows two main entities resembling the dictionary definition of God: the first God is described in *between* Genesis 1:1 and Genesis 2:25; only fifty-six verses out of the overall total verses (23,214) of the Old Testament describe who or what this God is. In brief, this God is loosely described as the God of planet Earth. The entire fifty-six verses can simply be summed up as worthy of modern-day kindergarten rhyme. Nothing short of sheer fantasy!

The other entity which shows up is the God of Israel, a sectarian or an ethnic God solely for Jews or Israelites from Genesis Chapter 3 and throughout the remaining thirty-eight books of the Old Testaments. Consequently, continued presentation of each book's review beyond the book of Judges (the 7th.) is unnecessary for our present discourses.

In addition to the above observations, some authors of the remaining thirty-two books of the old Testaments; at least five of them, (Ruth, Kings, Ecclesiastes, Isaiah, and Hebrews) deliberately add events which occurred after Christ, as opposed to events before Christ birth; plus the apparent needlessness of repetitions of identical Chapters and verses, as for instance 2 King 19 and Isaiah 37; more than 98 per cent exactly the same.

The importance of this deliberate neglect will come out when we discuss Christology in the New Testament era.

Thus Old Testament failed to give logical explanations of what constitutes a cosmic God. We now go to the other half of the Holy Bible, the New Testament narratives.

The twenty-seven books of the New Testament begin with the book of Matthew, and as usual the author's name and date of publication are missing. Based on its content, it is estimated to have been written about sixty to seventy years after the death of Christ. Again by its contents and history, the author did not ever meet Christ in person. The author is therefore reporting what he or she heard or read about Jesus and events during Christ era. We are informed in Matthew 9: 9; that 'Jesus was passing by, he saw a man seated at the tax desk, called Matthew, and said to him, "Follow me". And he got up and followed him.' When the gospel then lists Jesus's twelve chosen pupils, it specifies Matthew *as the man who made his living out of tax collection* (Matt. 10, 3). There is evidence that this disciple, Matthew, is not the author of this book.

It is also essential to note that there is no evidence whatsoever (theologically and historically) that someone by the name of Matthew had a revelation or a dream from God to write the book of Matthew. There is absolutely no supporting information that the spirit of Jesus or an angel from God dictated Matthew's gospel to anyone for publication. None ever happened. No proof.

The Gospel of Matthew tells a story that Jesus is the promised redeemer, the one through whom God fulfilled the promise he made to his people (Israelites) in the Old Testament.

Note that Matthew's good news is not for Jewish people only, but also for the world. It is not quite certain though whether this author's 'world' was the same narrow Jewish world perceived by the Old Testament authors, the rainbow end world. The book is, however, meticulously arranged.

It might interest you to know that some biblical scholars claim that there is recorded evidence that the four main Gospels under review now were written between fourth and fifth century AD.

It begins with the birth of Jesus and description of his baptism; his temptations and his ministry of preaching, teaching, healing and miracle he made in Galilee. The author records Jesus' journey from Galilee to Jerusalem, and the events of Jesus 'last week, culminating in his crucifixion and resurrection.

The author presents Jesus as a great teacher with God's power to talk about God's law, God's kingdom and how to get there. The book talks about destiny of man as the heavenly kingdom, and privileges and duties of citizens of heaven. The conclusion is focused on the death, resurrection and appearances of the Lord.

Matthew 1:1; 'This is the list of the ancestors of Jesus Christ, a descendant of David, who was a descendant of Abraham.'

Matthew 1. 17; 'So then, there were 14 generations from Abraham to David, and 14 from David to the exile in Babylon, and 14 from then to the birth of the Messiah.'

This makes a total of forty-two generations. But for clear understanding of this author, we must start from the following:

Verse 22: 'Now all this happened in order to make what the Lord has said through the prophet' (Isaiah) come true. 'A virgin will become pregnant and have a son, and he will be called Emmanuel' (which means 'God is with us').'

But, here is what Isaiah says, *Isaiah 7. 14;* Well, then, the Lord himself will give you a sign: a young woman who is pregnant will have a son and will name him 'Immanuel'. The Hebrew translation refers to any young woman of marriageable age, but definitely not a virgin. The author of book of Matthew uses Greek translation: Matthew 2. 4-6; King Herod called together all the chief priests and the teachers of the law and asked them, 'Where will the Messiah be born? In the town of Bethlehem in Judea, they answered.'

A Jewish prophet called Isaiah is claimed to have predicted about 900 years before Christ was born that a male Jewish child will be born to become the messiah who will save Jews from their sins. And in order to know who, when, and where that man is born, three prophesized conditions must be fulfilled; 1. The child would be from *David ancestry*, 2. A *virgin woman* would give birth to that person, and the male child would be born, 3; *in the town of Bethlehem*, Judea; *Micah 5, 2*

The author of book of Matthew is henceforth proving that the above three prophesies have come true, hence Jesus the Christ who he is about to narrate is unmistakably real.

1) In chapter 1.1-17; the author referred back to (14 x 3 = 42) forty-two generations from Abraham through David to Joseph genealogy and concluded that yes *Joseph* had blood link with David, hence the first prophesy is fulfilled. Note again, *not Mary's genealogy* but Joseph's genealogy.

2) In *Chapter 2.1*; '*Jesus was born in the town of Bethlehem, Judea during the time when Herod was king.*' Here again the second prophesy by Isaiah is deemed fulfilled according to the Gospel of Matthew: although we are informed by the same author that Jesus' home town is Nazareth.

3) In Chapter 1.18; '*this is how the birth of Jesus Christ took place. His mother Mary was engaged to Joseph, but before they were married, she found out that she was to have a baby by the Holy Spirit*'; verse 20-2: '*For it is by the Holy Spirit that she has conceived. She will have a son and you will call him Jesus, because he will save his people from their sins*'.

By the above three citations, Isaiah's three major predictions have been realized according to book of Matthew.

Matthew 1.18-23; '*This is how the birth of Jesus Christ took place. His mother Mary was engaged to Joseph, but before they were married, he found out that she was going to have a baby by the Holy Spirit. Joseph was a man who always did what was right, but he did not want to disgrace Mary publicly: so he made plans to break the engagement privately. While he was thinking about this, an Angel of the Lord appeared to him in a dream and said, Joseph, descendant of David do not be afraid to take Mary to be your wife, for it is by the Holy Spirit that she has conceived. She will have a son and you will name him Jesus, because he will save his people from their sins.*'

Needless to say the first two chapters are full of riddles, but they are not surprising if the reader reflects on our previous discourses of the Old Testament, the concerns of the chief architect of the biblical concept, the Synoptic view of Bishop Irenaeus of the 2nd AD; for instance, if Joseph was the surrogate father of Jesus, what will be the relevance of Joseph's blood genealogy as far as Jesus is concerned. Joseph and Mary did not have sexual intercourse before Mary got pregnant by the Holy Spirit; therefore Jesus had zero blood links with Joseph. The author's reference to forty-two generations from Abraham through David to Joseph as true evidence of Isaiah's prophesied virgin birth is as childish as the Genesis' narration of the cosmic creator. Pregnancy is possible only through interaction of female and male genitalia or male sperm and female egg fertilisation. These are the only two natural tools required for the creation of human beings. And there are no logical explanations why the supposed Genesis' God invented these organs. Some defenders of this biblical hypothesis often ask such questions as, who made these sex organs. They ask because they fail to accept the natural consequences of sexual interaction of males and females.

The flip side of understanding Jesus birth is for the author to state openly that the over seventy year-old Joseph had sexual intercourse with the under-sixteen-year infant Mary before she became pregnant. This will make the forty-two genealogy analysis by Matthew acceptable and significant.

Indeed, *Genesis 3. 20* clearly confirm this, that *'woman is the mother of all human beings,'* the universal factory or manufacturing plant responsible for creating people is the woman's womb. The only tools required in this factory is male and female sex organs coming in contact with each other. Everything else is mere philosophy. Both the author and you the reader are testimonies to this fact.

With regards to the place where Jesus would be born, the high priests said Bethlehem, but *Matthew 2. 23* are telling us that Jesus is from Nazareth, a province in Galilee as according to prophesies.

There appear to be another 'Da Vinci Code' hidden in prophesies written by the author of not only Matthew but also the remaining three other selected witnesses; Mark, Luke and John. Let us read again *Matthew 1. 22*; *'Now all these happened in order to make what the Lord had said through the prophet come true'*.

These so-called witnesses are telling readers in plain simple words *that something need be done to make all the instructions from the Lord to the Prophets become a reality.* The Virgin birth, David ancestry, Bethlehem birth place, and all prophesies *must* be *made* to *happen* so that the biblical god will be believed to be the God who actually created heaven and earth; hence he is supreme and he knows all future events before they occur. By implications then, it matters not the least whether to lie or to fabricate event occurrence, as long as what is predicted in the ancient scriptures is realize by any means possible, the end will necessarily justify the means. This Machiavellian code is undoubtedly secretly embedded in all biblical prophesies and dreams and they are purposely designed by ancient theologians, particularly, founders of Christianity to mesmerize inarticulate followers.

Still on the Jesus' mother narrative in the book of Matthew: it would have been thought that an important character such as Mary, the mother of Christ, (*Theotokos of Kyrios*) herself is mentioned only five times during infancy and only on one of these occasions *(Matt. 13. 55)* did the author make reference to Mary during *her adult* years. The remaining occasions are *Matthew 1.16, 18, 20; 2.11; 12, 46-50* and Mary is merely mentioned as 'mother of Jesus', although the name Christ and Jesus Christ is mentioned 325 times in Matthew's Gospel alone. Where is Jesus' mother's biography, for instance, her parents in the Holy Bible? Do we have to search through Gnostic scriptures to know that they are Anne and Joasham, plus the circumstances leading to Mary's birth: how she was brought up and by whom, and why as a twelve year-old Jewish girl, she had to marry Joseph, a widower five times older than her? We should not minimize Mary's significance if her son Jesus Christ is the one who transformed after death into the universal God. In effect those of you who see Christ as God must also see Mary as the mother of God. If so, then 'God' is man-made just like all of us. Just think about it!

Matthew 2. 1; 'Jesus was born in the town of Bethlehem in Judaea, during the time Herod was king.'

Matthew 3, 13-17;

> At that time Jesus arrived from Galilee and came to John at the Jordan to be baptized by him. But John tried to make him change his mind. I ought to be baptized by you, John said, and yet you have come to me! But Jesus answered him; let it be so for now, for in this way we shall do all God requires. So John agreed. As soon as Jesus was baptized, he came up out of the water. Then heaven was opened to him, and he saw the Spirit of God coming down like a dove and alighting on him. Then a voice said from heaven, 'this is my own dear son, with whom I am pleased.

Matthew 4, 14; 'This was done to make what the prophet Isaiah had said come true.'

The author of Matthew is telling us when and where Jesus was born. *Suddenly* Jesus is a full-grown matured man. He travels to Galilee to be baptized by John the Baptist. Jesus performs many miracles. The *Devil takes Jesus* to Jerusalem, the holy city, and sets him up with several temptations until angles came to rescue him. First question is how did the devil convey Jesus from Galilee to Jerusalem, was it by foot, by donkey or by spiritual flight? I know Christians would speculate the trip to be by spiritual means. But how possible can the devil control or tempt a super human being like Jesus who was not only conceived by the Holy Spirit, but also represents the true son of the Heavenly Father; the commander-chief of all powers. The devil took Jesus to no other place but to the holy city of Jerusalem. Where were the so-called angels when this mischief was going on? Why must God allow these so-called several temptations on his own begotten son? Although after baptism, heaven was opened to him, and he saw the spirit of God coming down like a dove and alighting on him. In joy, God pronounced him as his own dear son. Unbelievable faith!

Up to this point, would the story of Jesus not be complete if the author told readers the life of Jesus beyond baby Jesus? What and where he was when he was between one and eleven years old; or between twelve and nineteen years old, and indeed twenty to thirty years of his life? If these crucial periods of a human being's life are concealed for whatever reason and readers are informed that this character has had an encounter with a 'Heavenly' God, then there is every good reason to suspect a malicious hidden agenda which would include forgery, misinformation, or misrepresentation among all the characters involved; Mary, Joseph, John the Baptist, Jesus, and the so-called spirit Gabriel which arrived from heaven to impregnate a real human female.

As a reminder, it is very important to always reflect on the ultimate purpose of this discourse, which is, *a search for a Cosmic God;* does this particular character, Jesus, qualifies to be the inventor of Chinese, Indians, Egyptians, Eskimos, Pigmies, Europeans, Africans, Arabs and the rest of mankind? How can any rational human being believe this preposterous theology to be valid without any challenge? How?

The only God mentioned from chapter one to ten by this author is the spiritual God in heaven who is often referred to as Lord. The Ten Commandments of the Jewish God and the Laws of Moses are often cited in Jesus' sermons. He nevertheless renounces some of God of Israel regulations. For instance Jesus said according to *Matthew 5, 38-40;* *'You have heard that it was said, 'An eye for an eye, a tooth for a tooth. But now I tell you: do not take revenge on someone who wrongs you. If anyone slaps you on the right cheek, let him slap you on the left cheek too. And if someone takes you to court to sue you for your shirt, let him have your coat as well';* plus a few more don'ts. But in *verses 43 to 48* Jesus talks about the Perfect Father in Heaven, which resembles the book of Genesis godly character; the universe' designer.

In his teachings about how to pray, Jesus said:

Matthew 6, 8-13; *'Your Father already knows what you need before you ask him. This then is how you should pray: 'Our Father in heaven: May your holy name be honored: may your Kingdom come; may your will be done on earth as it is in heaven. Give us today the food we need. Forgive us the wrongs we have done, as we forgive wrongs that others have done to us. Do not bring us to hard testing, but keep us safe from the Evil One.'* Here the author has either lost his chain of thought already commenced or he is very confused! If the Heavenly Father already knows your needs, why the hell should the author teach us how to or why we should pray in the first place. Please read the first sentence again. This is absurd. Why can't the author simply state that God in Genesis (the universe creator) is different from both the Perfect Father in Heaven, the God of Israel and the Human God (the Christ). All the Gods in the Holy Bible have been testing humans since his so-called creation and this includes Adam and Eve, why should this author include the last sentence in the Lords' prayers (Do not bring us into hard testing). God knows humans detest temptations, yet he God was the first tempting governor on this planet.

Chapter 7, 12: *'Do for others what you want them do for you: this is the meaning of the Law of Moses and of the teaching of the prophets.'* And this is the same character Moses who murdered thousands of innocent people by the command of God of Israel during the Exodus. His laws are now seen as global laws. Is there any wonder if stronger cultures rise against weaker cultures; or imperial powers colonizing weaker nations? Who is the optimum beneficiary of exporting venoms of Moses to cultures which have survived for centuries in peace, on their own kind of Moses[s], and without the Laws of Moses?

The author of Matthew narrates the works of Jesus up to *chapter 20*, as an adult moral teacher, miracles performer and one who knows and understands Jewish traditions, prophets of the past, plus the nature of Heaven. But at the same time he behaves very often as if he was the Heavenly Father himself even before his alleged crucifixion and resurrection. The author at the same time tries to show Jesus as a real human being with natural siblings such as James, Joseph, Simon, and Judas as brothers; *Matthew 13, 55.*

Matthew 9, 6; 'I will prove to you that the son of Man has authority on earth to forgive sins.' Where did the author get this information from? He was not there when the statement was made. Historically the book of Matthew was written over sixty years after Jesus' death. The saying is therefore as a hearsay as everything else in the Bible, the statement is written in Quotation to give the impression that it was tape-recorded or copied from a written record, but it wasn't. Do we have to believe it without a question of its authenticity? Well, I don't think so!

Matthew 11, 2; 'When John the Baptist heard in prison about things that Christ was doing,' he sent some of his disciples to him; *'Tell us', they asked Jesus, 'are you the one John said was going to come, or should we expect someone else?'* Of all the 28 chapters in the book of Matthew, the name or title Christ is used only three times (as above and Chapter 1; 1, 18). It is about the baby Jesus all the way through. These little anomalies may be ignored as irrelevant by Christians, but they have serious implications about a human being who later become the cosmic god according to the infamous Holy Bible. Matthew appears to be unaware of Christ's preexistence before creation as prominently propagated by John and Paul. If he did, he would have applied this other name of Jesus, [the Christ], more often as John did to enhance Jesus spirituality in the book of Matthew. The book is surprisingly taken as the earliest book among the four main witnesses, and it hardly used the title Christ in its narratives. Rather it is the last witness, John, written about 100 years later which, again surprisingly, concentrates on Christ's title.

Where did the authors source their information? It would have been thought that the earlier writers would have been informed better and possessed more historical details than the later ones. But here in the Holy Bible it is the opposite.

Matthew 13, 55-57; 'Isn't he the carpenter's son? Isn't Mary his mother, and aren't James, Joseph, Simon, and Judas his brothers? Aren't all his sisters living here? Where did he get all this knowledge? And so they rejected him. Jesus said to them, 'A prophet is respected everywhere, except in his home town and by his own family.' From Matthew 21 onwards, Jesus had transformed into the prophesized 'symbol' portrayed in the book of Isaiah. Yes, a prophet will necessarily be very well known by his own family members and his own townsmen better than any others. All of us could be better judges of the prophet Jesus if these authors told the full news about Jesus rather than just his 'good news'. Where is the rest of his life story, [his bad news]?

And what does Jesus mean when he used the word family? Is he referring to his brothers, sisters, mother, and father; or his wife and children; or his religious followers; or a race of Jewish people? The word family has at least eight literal meanings.

According to this author's narrative, there is very little difference between the nature of Jesus in the New Testament and the nature of Moses in the Old Testament: for instance, the birth of both caused Pharaoh of Egypt and Emperor Tiberius of Roman Empire respectfully to hunt to kill all new born males. Moses was formed or created purposely to be the leading liberator of Enslaved Jews from Egypt to the Promised Land of Canaan as commanded by God of Israel: so was Jesus also formed or created by the same God, only this time the name is the Heavenly Father (not the Jewish God) to serve as the Messiah according to ancient prophesies to die for the sins of mankind.

Moses received instructions from the 'Tent of Gods presence' to go to the top of a Mountain to receive the Ten Commandments from God. While Jesus had spiritual revelations as the only begotten son of God who conducted sermons on mountain tops and to die for the sins of humanity.

Both are Jews from the same genealogy and died of similar mysterious circumstances.

This book draws a clear line between the Old and the New Israel, the traditional circumcised Judaic Israelites who live by the Law of Moses and Judaic scriptures on one side and the Jews who followed the deviant pagan Christ as dictated by the Heavenly father through his son Jesus Christ.

In Matthew 16: 20, 'Jesus orders his disciples not to tell anyone that he was the Messiah.' What is he afraid of?

Suddenly, a plot to kill Jesus is revealed by Jesus himself to his disciples. *Peter 26: 1-4, 'When Jesus had finished teaching all these things, he said to his disciples;' in two days as you know, it will be the Passover Festival, and the Son of Man will be handed over to be crucified. 'Then the chief priests and the elders met together in the palace of Caiaphas, the High Priest, and made plans to arrest Jesus secretly and put him to death.'* Jesus knew every detail of the plot and its definite outcome, and this makes the so-called plot a lie.

But before his arrest, according to Matthew's author, he went to Bethany at the house of Simon, and

Matthew 26, 7-12;

> While he was eating, [a woman] came to him with an alabaster jar filled with an expensive perfume, which she poured on his head. The disciple[s] saw this and became angry. Why all this waste? [They] asked. This perfume could have been sold for a large amount

and the money given to the poor!' Jesus knew what they were saying, so he said to them. Why are you bothering this woman? It is a fine and beautiful thing that she has done for me. You will always have poor people with you, but you will not always have me. What she did was to pour this perfume on my body to get me ready for burial'.

After this event, Judas agrees to betray Jesus; an intricate malicious deal to make ancient prophecies become a reality.

By the tone of the alleged plot between the Council of Rabbis and Judas Iscariot, Jesus appears to be a major accomplice to the case of his own alleged crucifixion; just to make ancient prophecies come true. Thus the plot, his crucifixion and resurrection are untrue.

Matthew 26: 14-16; 'Then one of the twelve disciples—the one named Judas Iscariot—went to the chief priests and asked,' What will you give if I betray Jesus to you? 'They counted out thirty silver coins and gave them to him. From then on Judas was looking for a good chance to hand Jesus over to them.' At the Passover meal with his disciples, Jesus said to his disciples, verse *18 'My hour has come; verse 21; I tell you one of you will betray me. One who dips his bread in the dish with me will betray me. Verse 23; The Son of Man will die as the scriptures say he will.'* Then at the Last Supper Jesus gave bread and drinks to his disciples said, verse *28; 'this is my blood, which seals God's covenant, my blood poured out for many for the forgiveness of sins.' Verse 31-32; 'Then Jesus said to them, this very night all of you will run away and leave me, for the scripture says; God will kill the shepherd and the sheep of the flock will be scattered. But after I am raised to life, I will go to Galilee ahead of you.'*

Matthew 26: 45; 'Then he returned to the disciples and said, Are you still sleeping and resting? Look the hour has come for the son of man to be handed over to the power of sinful men. Get up, let us go. Look, here is the man who is betraying me!' Verse 48-50; 'The traitor had given the crowd a signal: the man I kiss is the one you want. Arrest him! Judas went straight to Jesus and said, Peace be with you, Teacher, and kissed him. Jesus answered. Be quick about it, friend!'

Verse 53; 'Don't you know that I could call on my Father for help, and at once he would send me more than twelve armies of angels?' But in that case, how could the Scripture come true, which says that this is what must happen?'

Of course the Scriptures must come true regardless; otherwise human instinct would have prompted Jesus that his life was in danger hence necessary precautions must be taken to avert the impending danger. I am sure any of his disciples or indeed any of us would have behaved contrary to what Jesus did. The scriptures say *'this is what must happen'*. Jesus stood there to be kissed by Judas, after saying 'Peace be with you' and Jesus responded 'Be quick, about it friend!' If these sayings by Jesus are true according to the author, the Jesus story from his conception by Mary, his missing biography, his magical performances, his links with the mythical heavenly father to his death; give the obvious impression of a repeated fictional plays from Genesis' God to Moses and the Exodus, to God of Israel and the rest; all fictions, nothing but fictitious ancient fables; believable only by primitive and out-of-focused minds. I do not intend to be offensive here, but the alleged behaviour of Jesus regarding his assassination plot is irrational, if he was indeed as human as you and I.

Verse 56-57;'But all this has happen in order to make what the prophets wrote in the Scriptures come true. Then all the disciples left him and run away.'

Those who had arrested Jesus took him to the house of Caiaphas, the High Priest, where the teachers of the Law and the elders had gathered together.

Verse 59-66;

> The chief priest and the whole Council tried to find some false evidence against Jesus to put him to death; but they could not find any: even though many people came forward and told lies about him. Finally two men

stepped up and said, this man said, I am able to tear down God's Temple and three days later build it up again. The High Priest stood up and said to Jesus, have you any answer to give to this accusation against you? But Jesus kept quite. Again the High Priest spoke to him; in the name of the living God I put you on oath to tell us if you are the Messiah, the Son of God. Jesus answered. 'So you say. But I tell all of you, from this time on you will see the Son of Man sitting on the right of the Almighty and coming on the clouds of heaven!' At this the high priest tore his clothes and said, Blasphemy! We don't need any more witnesses! You have just heard his blasphemy! What do you think? They answered; He is guilty and must die.

Please be reminded once again at this point that we are trying to find out from the book of Matthew, as written in the New Testament of the Holy Bible, if it contains any information that can be used to support the validity of a real human being who became a God for the whole world. The Jews and their elders who lived in Jerusalem at the time were aware of their ancient prophesies, Laws of God of Israel, and Moses Laws; only a few Jews in Jewish communities have also heard of Jesus claims of being the prophesized Son of God.

Again they have also heard that Jesus's sermons mostly condemned the usual rules, traditions and regulations in the Jewish communities. Hence his own tribesmen did not like him.

Isaiah 29: 13; 'The Lord said, these people claim to worship me, but their words are meaningless, and their hearts are somewhere else. Their religion is nothing but human rules and traditions, which they have simply memorized.' Evidently, none other than a fraction of Jewish people at the time, including Jesus own disciples did not accept his claims. Hence at the rabbinic council hearing or trial, it was expected of Jesus to prove beyond all reasonable doubts that he was really what he claimed to be, but he failed. And he failed simply because he merely wanted *'to make the scripture and the ancient prophesies come true' through Jesus, himself:* notice that Jesus used this very clause many times.

Matthew 27, 1-2; 'Early in the morning, all the chief priests and the elders made their plans against Jesus to put him to death. They put him in chains led him off, and handed him over to Pilate, the Roman governor.'

Verse 9: 'then what the prophet Jeremiah had said came true'.
Verse 15-26;

> At every Passover Festival the Roman governor was in the habit of setting free any one the crowd ask for. At that time there was a well known prisoner named Jesus Barabbas. So when the crowd gathered, Pilate asked them, which one you want me to set free for you. Jesus Barabbas or Jesus called the Messiah? He knew very well that the Jewish authorities had handed Jesus over to him because they were jealous. While Pilate was sitting in the judgement hall, his wife sent him a message: have nothing to do with that innocent man, because in a dream last night I suffered much on account of him. The chief priest and the elders persuaded the crowd to set Barabbas free and have Jesus put to death. But Pilate asked the crowd, which one of these two do you want me set free? Barabbas! They answered.
>
> 'What then shall I do with Jesus called the Messiah? Pilate asked them. Crucify him! They answered. But Pilate asked, what crime has he committed? Then they started shouting on top of their voices: crucify him! I am not responsible for the death of this man! This is your doing!
>
> Then Pilate set Barabbas free for them: and after he had Jesus whipped, he handed him over to be crucified.

The one thing that makes Roman Empire unique in history is the respect of their Laws. Jurisprudence was a major concern of all the Emperors and Roman citizens. You are judged guilty when proven beyond all reasonable doubts. The Governor Pontius Pilate had no legal authority under Roman Laws at the time to hand Jesus over to the Jewish chief priests to be crucified. It is not possible that a Roman Governor will deliberately abuse the law by not giving Jesus due process of the law; to be proven guilty after a fair trial. What the Gospels say is incorrect and a lie. It is an absolute impossibility for a Roman Governor to flout prosecutorial laws at the time!

About thirty to forty years after Jesus' crucifixion, the Apostle Paul was under the same predicament in Jerusalem, and he was allowed to appeal to the Emperor in person. In fact he was given legal assistance and official escort from Jerusalem to Rome, Italy because of justice: (*Acts 25: 11-12*).

The story of Jesus as narrated by this author and the other supposed witnesses cannot be true. History and Common sense does not support it, the then Jewish tradition does not support it, the Laws of Moses does not support it, and God of Israel Laws does not support it. The entire story is as fictitious as that of the Genesis' God (the Creator) and the Jewish God of the Pentateuch; they are all blatant lies.

One other interesting issue to note in the above quotation is (*Matt. 27, 16*); the name of the famous prisoner 'Jesus Barabbas'. In Hebrew, Barabbas means 'son of Rabbi'. Do we take this prisoner to be the 'son of the Messiah,' Rabbi Jesus Christ' own son? Readers of the Holy Bible are informed that all recognized Jewish religious leaders, including Jesus Christ, are called Rabbis. Why was it not the life of any other prisoner at the time, but 'Jesus Barabbas' and that of 'Jesus Christ' as a bargain chip? Is it a mere coincidence or that Barabbas was in fact the son of Jesus Christ? At the time of his kangaroo trial by Governor Pilate, Jesus Christ was about thirty-three years old and capable of having a son seventeen years old and above. It was and still is a Jewish tradition for every Rabbi to be a married man with children or at least a child.

Jesus Christ was always addressed by his Jewish followers as Rabbi and he cannot be the only exception in rabbinic history to be honored a Rabbi or a Priest with no wife and no children.

The fact that sex life of Christ is hidden, and out of popular history books, gives impression that the man probably had a nasty or unconventional attitude to sex which is not worth publicity.

Verse 57-61;

> When it was evening a rich man from Arimathea arrived. His name was Joseph, and he also was a disciple of Jesus. He went into the presence of Pilate and asked for the body of Jesus. Pilate gave orders for the body of Jesus to be given to Joseph. So Joseph took it wrapped it in a new linen sheet, and place it in his own tomb, which he has just recently dug out of solid rock. Then he rolled a large stone across the entrance to the tomb and went away. Mary Magdalene and the other Mary were there, facing the tomb.'

Note that there were only three people at the tomb site and Mary, the mother of Jesus was not included. These three were Joseph of Arimathea, Mary Magdalene, and the other Mary. This other Mary could be either one of Jesus's sisters, Mark 13: 56, or the Mary of Bethany, but the author would have mentioned it; or the twelve year-old girl who Jesus addressed as [my daughter] in Mark 5: 34.

The Resurrection: *Chapter 28: 1-10;*

> After the Sabbath, as Sunday morning was dawning, Mary Magdalene and the other Mary went to look at the tomb. Suddenly there was a violent earthquake; an angel of the Lord came down from Heaven, rolled the stone away, and sat on it. His appearance was like

> lightening, and his clothes were white as snow. The guards were so afraid that they trembled and became like dead men. The angel spoke to the women. You must not be afraid. I know you are looking for Jesus who was crucified. He is not here: he has been raised, just as he said. Come and see the place where he was lying. Go quickly now and tell his disciples. He has been raised from death, and now he is going to Galilee ahead of you; there you will see him. So they left the tomb in a hurry, afraid and yet filled with joy, and run to tell his disciples. Suddenly, Jesus met them and said; peace be with you . . . Go and tell my brothers to go to Galilee, and there they will see me.

In comparison, Matthew and Luke are completely different, regarding Jesus' resurrection. Which of them is the truth, or we are not supposed to reason on such issues just because the truth might wash out our faith?

Matthew 28: 10;

> The eleven disciples went to the hill in Galilee where Jesus had told them to go. When they saw him, they worshiped him, even though some of them doubted. Jesus drew near and said to them, 'I have been given all authority in heaven and on earth. Go then to all peoples everywhere and make them my disciples, baptize them in the name of the Father, the Son and the Holy Spirit, and teach them to obey everything I have commanded you. And I will be with you always, to the end of the ages.

Now please, read the above Bible quotation again. Doesn't it sound like the author is an eye witness of Jesus' alleged resurrection story? It reads as if he or she was there with the two Mary's, the tomb Guards and one angel. It is even more bizarre when you read the testimonies of Mark, Luke, and John; they too claim to be witnesses; Mark saw three women, the two Mary plus Salome, (not Joanna) and a young man in a white robe sitting: Luke also saw more than three women, the same two Mary, but third woman was Joanna and the *other women (no names)* with them plus two angles (not one angel); John however saw only Mary Magdalene plus two angles. His mother, Joanna, Salome and the others women were not seen on site according to the Gospel of John. No court of law will entertain any of these witnesses.

My main concern at this moment is the inconsistencies among these four carefully selected canonical gospels. Some defenseless Christians, the Apologist, try their best to offer defenses for these biblical anomalies. But if among the over 600 testimonies which confronted the First Council of Nicaea 292 years after Christ death, this is the best reliable evidence Christianity could offer to maintain the faith, then I see it as a phoney religion. If on the other hand the element of this religion is seen as universal crusader of public and private morality, which is completely divorced of divinity, I would join those who sing halleluiah everyday.

The use of biblical God and its sundries names have always been very effective tool to manipulate out-of-focused minds. A focused mind contains a dream and therefore has no room for manipulation, superstitions, spirituality or divinity. A focused mind concentrates on the inner faith (not divine faith) and convinces the self that all is possible as long as the brain is fully engaged on a goal to accomplish X, whatever your X might be without mentioning *God*. You may pray to the Genesis' God, the God of Israel, Jesus, Christ or Jesus Christ or the Heavenly Father or angels for assistance or help; truly, truly I tell you, you merely throw words into the air. Just *create* and *maintain* a defined dream, stay focused and make a move, and all help and assistance will be realized from within the self not from without the self. There is simply no god anywhere other than you. You are your god. Your brain is not a divine entity; it is your individual verifiable biological organ purposely for capturing, maintaining, and processing information for your personal benefit. If you fail to use it wisely for your benefit, others like theologians will manipulate it and use it for their benefit. It is written in the Gospel of Philip that 'Ignorance is the mother of all evils', and knowledge is freedom; 'if we know the truth, we shall find the fruits of the truth within us'. Not outside us. Not prayers.

During his lifetime, Jesus himself wondered [*Matt. 16, 13-16*] '*who the people say the Son of Man is?*' and the people of Caesarea Philippi for instance messed up; some said he was John the Baptist, others said Elijah or Jeremiah. He also asked his disciples the same question, '*Who do you say I am?;*' and Simon Peter answered '*you are the Messiah, the Son of the living God,*' but strangely, in *Matt. 16, 20*; '*Jesus ordered his disciples not to tell anyone that he was the Messiah.*' Yet after his resurrection *Matthew 28, 18-20*; Jesus tells his disciples '*I have been given all authority in heaven and on earth. Go, then to all people everywhere and make them my disciple: baptize them in the name of the Father, the Son, and the Holy Spirit, and teach them to obey everything I have commanded you. And I will be with you always, to the end of the age.*'

After his crucifixion and resurrection according to the book of Matthew, Jesus became de facto controller of the universe; indeed the Omni cosmic God; almost identical to the God exhibited in book of Genesis, but of the same status as the God of Israel. Matthew however refused to call Jesus the 'Christ,' rather the title Messiah 'not' the Anointed, is assigned to him throughout. The multiple pictures of Jesus presented to us by the ancient Rabbinical Council of Bishops through the Bible, clearly demonstrates their desperate effort to make it appear real; that this world and all things herein is under the management of one God who must by all means be a Jew. Thus such questions as "who do you say Jesus Christ was?" cannot be answered to everyone's satisfaction. Because the surrounding circumstances from his conception to his birth, his name, his title, his missing biography, his works, and his crucifixion and to his resurrection makes the name Jesus Christ a fictional character. Every thing about the guy is fabricated. None of the stories, the scenes or episode resembles a real human being experience. And even if, we choose to define him as a spiritual entity, he becomes 'something' not someone, which exists only in our individual minds: the name and the title Jesus the Christ still becomes a mere perception.

A search in biblical literature and history give some clues to a better understanding of the character 'Jesus the Christ.' This will be explained fully after our review of the Book of John.

It must however be remembered again that our quest is a search for the Universal God. The important question which need be asked is whether the book of Matthew was able to show the reality of such an entity? The obvious reply is no! Readers are *enticed* with the supposed sayings and deeds, plus the circumstantial birth, death, and resurrection (in human form) of Jesus Christ as the Universal God.

There are too many question marks about Matthew's Jesus Christ that makes this man-made God unacceptable. He may be a Jewish cultural God for some Jews just like any ethnic cultural God existing in the various cultures of the world. *The search still goes on.*

ii.) The Book of Mark (Jesus Christ, Son of God)

The Gospel of Mark begins with the statement; 'this is the Good News about Jesus Christ, the Son of God.' As usual, the author's identity is missing. It is known however that the name Mark does not represent any person who knew or ever met Jesus Christ in person. Mark is only the title or the name of a book in the Holy Bible. Jesus is projected as a man of action and power.

His authority is seen in his teaching, in his power over demons, and in forgiving people's sins. He speaks of himself as the Son of Man who came to give his life to set people free from sin.

The book presents the story of Jesus in a straightforward, forceful way, with emphasis on what Jesus did, rather than on his words, teaching and his personality. After a brief initial remark about John the Baptist, and the baptism and temptations of Jesus, the author immediately takes up Jesus's ministry of healing and teaching. In his later years, followers of Jesus came to understand him better, but his opponents became more hostile. The book's closing chapters' reports of the events of Jesus' last week of earthly life, especially his crucifixion and resurrection. This story has Two or Three separate Endings: one by the author, the others are Two Older versions' Endings, which appear to have been written by some writers other than the first anonymous author.

Mark 1: 1-2; 'This is the Good News about Jesus Christ, the Son of God. It began as the prophet Isaiah had written;'

Why the Good News, what about the 'Other News', 'The not-so-good' News? The author is referring to only the best part of Jesus Christ story. There is a presumption of knowledge of the other side of the News. In other words the author is about to reveal *only* the information that is favorable to his agenda or the news that portrays Jesus as Holy and unblemished person. By implication therefore the author is saying that the public deserve to know only the good deeds of Jesus Christ.

Incidentally, just about fifty years after the book of Mark was written; and also about 100 years after Christ death, the book of John came out with a notation (John 21, 25); that there were many other things that Jesus Christ did. That if they all were to be written down, the whole world could not contain the number of books. John affirmed clearly that Jesus Christ' story is not just Good News but it also includes the opposite [BAD] news as well. It is very interesting that none of the so-called apostles or gospels is bold enough to testify to even a single fault of Jesus the Christ. Of course, he must be painted holy, pure faultless innocent Son of God. But I like to invite the reader to pulse for second and reflect for a moment if it is ever possible for a human being who was formed in a woman's womb and delivered from a woman's vagina and lived on this planet earth for thirty-three years could be innocent of a sin, as we are being told in the New Testament of the Holy Bible. Honestly, it is impossible; yet we are compelled to believe the Gospels as the holy truth.

Now, let us ponder on another clause in Mark's quotation; 'the Son of God'. In our previous book, 'Matthew' used the same term 'the Son of Man'. Literally the word Son refers to a human male child; while God represents a supernatural entity, according to Bible. God is also the inventor of the universe and everything under the sun. Hence by definition the clause means *the child of the maker of the world*. The term 'Son of God' is very ticklish, because we are told that the male child in question is biblically called Jesus Christ, who we are informed, was a meshed creature; partly human and partly divine. A twist of Plato's concept of man, as consisting of two parts; the physical visible human body and the invisible spiritual part: both belonging to different worlds, but coexists in the same person.

Obviously the human side of this creature (Jesus) is like all of us, but his divine element can only be viewed in our minds or through a thought process; the acquired total information stored in our brain organ, 'the mind' when processed can show us the nature of this Christ creature.

The Holy Bible carries a lot of clues about this man. To many, the invisible part is the soul or the Holy Ghost.

According to Bible, the divine Christ has had perpetual existence since creation. That is, the Christ who was the natural birth son of Mary was already alive on planet earth millions of years before Mary was born. This explains why in *Genesis 1, 26*; *'Then God said, and now* we *will make human beings; they will be like* us *and resemble* us*'*.

Thus by fusing the two elements together, the human side [Jesus] and the spiritual side [Christ], you practically invent a fictional character equal to God the Creator himself. If we can tune back to the book of Genesis through to Mark, and indeed to the end of book of Revelations, we will notice that all the authors attempt to introduce fictional characters, seemingly different, yet identical in their metamorphosis and resonance towards a utopian world with an imaginary human or a spirit that has no real existence.

Unfortunately the entire biblical books are based on extremely loose hypotheses when closely examined. For instance, if Jesus Christ is really the Son of God, who is Mary then; is she the Mother of God? What about Anne and Joachim, Mary's parents; and Jesus brothers and sisters; and the other siblings? And wouldn't there be two separate Gods at the time of the universe' creation, and also that the two plural pronouns [we, us] as used in Genesis would then become sensible? Again it would be a mistake to conclude that there were two gods in one entity at the time of creation; because one entity would not require such words as *we are* and *let us*. At a glance, the Editors of the Holy Bible appear to be good at inventions, but when critically reviewed none is perfect, because of the books' inherent numerous questionable invented characters and their misguided sayings.

Besides, these so-called sayings by God and Jesus and the rest which are written in quotes are not quotations sourced from the characters own recorded documents. All the sayings are evidently hearsays by later generations and therefore cannot be accepted as unquestionable truth. For instance, it is only *Mark 16: 6;* which states that *'Don't be alarmed. I know you are looking for Jesus of Nazareth who was crucified . . . '.* Where did the author of book of Mark get this quoted information from that the rest of the gospels new nothing about?

The Book of Mark contains only 16 chapters, but with two separate endings; an Old ending with 20 verses, but with verses 9 to 20 showing as a later day insertion. It appears that the author originally ended the last Chapter 16, at verse 10, but later generation of Bishops, possibly Irenaeus or Valentine, realized that other vital narratives were missing and need be added. The verses 9 to 11 for example deals with *Mary Magdalene and Jesus' resurrection*: verses 12 and 13 talks about *Jesus appearance to two Disciples*; verses 12 to 18 talks about *Jesus appearance to the eleven disciples after resurrection with instruction for them 'to go throughout the whole world and preach the gospel to all mankind';* then lastly verse 19 and 20 talks about the time when *Jesus is taken up to Heaven where he sat at the right side of God. The disciples went and preached everywhere, and that the Lord worked with them and proved that their preaching was true by the miracles that were performed*'.

The tone of the above additional verses when compared to the rest of the narratives is clearly modernized and different. It does not carry the Jesus' era choice of words; the First century parables and colloquial sentences. We can certainly conclude that the additional texts were inserted during Luther's time in the Middle Ages. Over a thousand years after Jesus' supposed death. Unity of ideas surrounding the character is still lacking, simply because all of them are inconsistent fabricated stories intended to reinforce Christians' faith in the gospel.

iii.) The Book of *Luke* (the Physician)
(Jesus Christ, the Son of God)

The Gospel of Luke presents Jesus as both the promised Savior of Israel and as the Savior of mankind. Luke records that Jesus was called by the Spirit of the Lord to preach the Good News to the poor. And this Gospel is filled with a concern for people with all kinds of need. The note of joy is also prominent in Luke, especially in the opening chapters that announce the coming of Jesus, and again at the conclusion when Jesus ascends to heaven. The story of the growth and spread of the Christian faith after the ascension of Jesus is told by the same author in the book of Acts.

Luke Chapter 1 to chapter 2, 52; deals with general introduction of the book and continues with the birth and childhood of John the Baptist and of Jesus. The ministry of John the Baptist and the baptism and temptation of Jesus are covered in chapters 3, 1-20; and 3, 21 and 4-13 respectively. There is extensive coverage of Jesus public ministry in Galilee between chapters 4, 14 and chapter 9.50; from Galilee to Jerusalem is also narrated in 9. 51-19.27. Jesus' earthly last week as spent in Jerusalem is covered in chapter 19, 28 to chapter 23, 56; his resurrection, appearances, and ascension to the Lord are narrated in chapter 24. 1-53.

Historically the third book of the New Testament (Luke) was written about seventy to eighty years after the death of Jesus Christ. It is not known who the real author is, although the authorship has been assigned to many including Luke himself who we are informed was a physician. There appear to be a consensus that the second century Bishop of Lyons (Irenaeus) was the chief Editor among others like Hippolytus, Origen, Justin Martyr, Clement, and Ignatius who conducted a search and compilation of all the ancient biblical texts. These original compilations went through several theological revolutions during and after the pioneering era of first and second centuries until the middle of fifteenth century when the present document took its present form to become the Holy Bible. It is worth being notified also that almost every generation attempt to modify the Gospels to befit its religious and social needs, hence the existence of several versions and multiple interpretations of the entire sixty-six current books.

Thus, if we measure the history of the Holy Bible and evaluate its content, every reader would need to maintain a serious *caveat* of human beings' fallible nature. Please be warned! What margin of error can the Bible readers attach to the Gospels? An answer to this of course will particularly depend on our personal judgmental capabilities, not faith; but to accept every biblical word in good faith, just for faith sake without forewarning is a grievous human risk. The authors of the books in the Holy Bible and the people who put the hearsay ideas into logical readable form were all as humans as yourself. Some claim to be God-appointed in the same way as you and I can be self-appointed by *our conscience*. If self-confessed murderers like the Apostle Paul (*Act 9, 21: 22, 20*) can become an Apostle or an Evangelist; and if a human being like Jesus can become a God (the Jesus who refused to be called 'good' because he knew God is the only one who deserves to be called good *Mark 10; 18*), why can't you and I? Even Jesus himself knew that he was not perfect.

Now let us find out if the author of Luke has a convincing proof of a possible cosmic God; the designer of all humans.

Luke C*hapter 1, 1-4; 'Many people have done their best to write a report of the things that have taken place among us. They wrote that we have been told by those who saw these things from the beginning and who claimed the message.'*

And so your Excellency, I have carefully studied all these matters from their beginning, I thought it would be good to write an orderly account for you. I do this so that you will know the full truth about everything which you have been taught.' Luke, the doctor has cautioned his readers that he is not an eye witness of what he is about to tell us. He has however carefully studied the content of what he has read and feels confident that his Excellency, Theophilus, will find his narratives truthful and useful.

Luke 1: 11; 'An angel of the Lord appeared to him, standing on the right side of the altar where the incense was burnt. When Zachariah saw him he felt afraid, but the angel said to him: 'Don't be afraid, Zachariah! God has heard your prayers and your wife Elizabeth will bear you a son. You are to name him John'.

Since the time of Abraham, all heroes in the Holy Bible are either mysterious conceived or miraculously given birth to. And either both parents or one of them, or their grandparents, are above childbearing Menopausal age or grossly underage females, or barrens.

Verse 15-17; 'He will be a great man in the Lord's sight. He must not drink any wine or strong drink. From his very birth he will be filled with Holy Spirit, and he will bring back many of the people of Israel to the Lord their God. He will go ahead of the Lord strong and mighty like the prophet Elijah. He will bring fathers and children together again; he will turn disobedient people back to the way of thinking of the righteous; he will get the Lord's people ready for him.'

The author is referring to the Lord God of Israel, not the Genesis God or the Heavenly Father. It is very tempting for readers to link these gods as if they are one and the same God, but the scriptures in the sixty-six biblical books do not say so. The Israeli

God says he is only for his people [Jews] or the children of Israel. If you are not a circumcised Jew', but a Gentile or a person of any culture, you do not belong to the God of Israel. Please, read carefully the above quotation in italics.

Verse 18; 'Zachariah said to the angel, how shall I know if this is so? I am an old man and my wife is old also.'

'I am Gabriel, the angel answered. I stand in the presence of the God who sent me to speak to you and tell you this good news.' The Jewish God sends his messenger angel Gabriel to deliver a message to John the Baptist' father, Zachariah, that his age and that of his wife Elizabeth are not important; they are the chosen couple to bring forth this future great Jewish hero [John the Baptist]

Verses 24-25: *Some time later his wife Elizabeth became pregnant and did not leave the house for five months. Now at last the Lord has helped me, she said. He has taken away my public disgrace!* Elizabeth is five months pregnant and very happy and no more a disgraceful barren. The baby's movement in her abdomen when Mary visited her with her own miraculous pregnancy story [a virgin, yet pregnant] gives clear indication of how great God of Israel is. The fetus in Elizabeth's womb started jubilating over Mary's pregnancy news. Besides the above information about John in the mother's womb, nothing is written anywhere in the Bible of when, how, and where John the Baptist was born, how he grew up as a child, as a teenager and the age at which he died.

Matthew, Mark, Luke and John the Apostle testify about his preaching and baptizing famous people like Jesus Christ and how he was beheaded in prison, nothing else about him is known to posterity. He is merely portrayed as a metaphor of Jesus Christ.

Luke 1: 26-31: The birth of Jesus is announced. *'In the sixth month of Elizabeth's pregnancy God sent the angel Gabriel to a town in Galilee named Nazareth. He had a message for a girl promised in marriage to a man named Joseph who was a descendant of King David. The girl's name was Mary. The angel came to her and said,' Peace be with you! 'The Lord is with you and you are greatly blessed. Mary was deeply troubled by the angel's message and she wondered what his words meant. The angel said to her don't be afraid, Mary; God has been gracious to you. You will become pregnant and give birth to a son, and you will name him Jesus.'*

This is the beginning of a human God; a real visible human being with bones, flesh and blood; *Jesus Christ*; who later transformed as God the creator. How true is this? We will find out soon.

Verse 32; 'He will be great and will be called the son of the most highest, God. The Lord will make him a king as his ancestor David was, and he will be king of the descendants of Jacob for ever; his kingdom will never end!'

Verse 34-35; 'Mary said to the angel, I am a virgin. How then can this be? The angel answered; the Holy Spirit will come on you, and God's power will rest upon you. For this reason the holy child will be called the Son of God.'

Verse 36-38; 'Remember your relative Elizabeth, it is said that she cannot have children, but she herself is now six months pregnant even though she is very old; for there is nothing that God cannot do. I am the Lord's servant, Mary said; may it happen to me as you have said. And the angel left her.'

Verse 67-68; 'John's father Zachariah was filled with the Holy Spirit, and he spoke Gods prophesy; Let us praise the Lord, the God of Israel'

Luke 1, 78; 'Our God is merciful and tender.'

God of Israel is not merciful and tender; isn't he the same God of Israel who commanded Moses and Joshua to kill innocent citizens of Jabesh, Amorite, Ai, Moab, Midian, Evi, Rekem, Zur, Hur, and Reba. *Numbers 30, 17-18; Numbers 31, 1-20; Numbers 14, 35-38. Deuteronomy 12, 32; 'Do everything that I have commanded you, do not add anything to it or take anything from it.' Deuteronomy 13, 8-9, 15-18. Deuteronomy 20, 10-18. Deuteronomy 32, 39-43. Deuteronomy 33, 26; 'No God is like your God.' Deuteronomy 32, 9; 'He assigned to each nation a god, but Jacobs descendants he choose for himself' Joshua 6, 17-21; Joshua Chapter 8, 1-29; Joshua 6, 16-21. Judges 23, 10-11; Judges 20, 23, 27-28; Joshua 11, 6-9.* In all the quotations above (from Numbers to Judges) none of the victims had done anything wrong, yet God of Israel commanded Moses and Joshua *to kill* them.

Now the Birth of Jesus according to the author of the book of Luke.

Luke :, 1-7;

> At that time Emperor Augustus ordered a census be taken throughout the Roman Empire. When this first census took place Quirinius was the governor of Syria. Everyone then went to register himself, each to his own town.
>
> Joseph went from the town of Nazareth in Galilee to the town of Bethlehem in Judea the birth place of King David. Joseph went there because he was a descendant of David. He went to register with Mary, who was promised in marriage to him. She was pregnant, and while they were in Bethlehem a time came for her to have her baby. She gave birth to her first son, wrapped him in strips of cloth and laid him in a manger; there was no room for them to stay in the inn.

Was Jesus a native of Galilee or a native of Judea? Well, may be the human Jesus *should* be a native of Nazareth in Galilee, but as a Christ, Jesus *must* be a native of Bethlehem in Judea so that Isaiah' prophesy will come true. The dual nature of this character is always displayed in the Gospels.

Luke 2: 11; 'This very day in David's town your savior was born; Christ the Lord.' Here the physician's precision is clear; that Christ, *not necessarily*, Jesus is born a citizen of Bethlehem, the home town of David as prophesized. It should be noted that we are searching for the cosmic creator (God). Luke is desperately trying to impress us that his character 'Christ' is capable of being the universe designer even though he was born on planet earth, *not* outside planet earth; and also not from any other ethnic group but of Jewish ancestry.

Luke 2: 22; 'A week later, when the time came for the baby to be circumcised, he was named Jesus, the name which the angel has given him before he had been conceived.'

Of course, as a true Jew, Jesus must be circumcised. Note that the author is not using Jesus' second name 'Christ,' because it would have been circumcision of a spiritual entity.

Luke 2: 42-43; 'When Jesus was twelve years old they (Mary and Joseph) went to the festival as usual. When the festival was over, they started back home (Nazareth), but the boy Jesus stayed in Jerusalem. His parents did not know this.'

This story cannot be true, because responsible adults do not behave like that. The author stops abruptly with the life stories of Jesus and rather continues with genealogy or the ancestors of Jesus. Did the author hear anything about Jesus when Jesus was between thirteen years old and thirty years old? Are we not entitled to know it if he or she knows it? An unfinished story such as this clearly tells readers who are not satisfied with half stories to search for the other untold part. A thirty year-old human being must have passed through twelve to twenty-nine years before reaching thirty; or are we to assume his lapsed years as part of Jesus's miracles; the incredible man who avoided teenage life and the entire 20s even though he lived to the age thirty-three years before he supposedly died.

Luke 3: 23; 'When Jesus began his work, he was about thirty years old. He was the son of Joseph who was the son of Heli.' How can we be convinced that this Jesus was a real human being?

Luke 3: 23-38; this author also recounts Jesus' genealogy, beginning with Joseph, father of Jesus (not Jesus' mother Mary), and ended the genealogy track with 'the son of Enosh, the son of Seth, the son of Adam the son of God;' note that the chain of blood link ends with Adam, the first male human which Genesis God created; many millions of years before David or Joseph were born. This author compressed the several millions of years into mere seventy-seven generations.

Between chapter 4 and chapter 21, Jesus' sermons, his miracles, his healings, and prophesies are narrated in greater details than Matthew and Mark. And everything that makes god a God is demonstrated by the author of Luke.

The author of Luke continues the narratives in the book of Acts to testify that even though he, Luke, admits he is not an eyewitness to the story he is telling us; he is nevertheless convinced, and we should be convinced too that his version of Jesus' story is nothing but the truth. I leave you the reader to judge for yourself. Remember though that the book of Luke was written about eighty years after Jesus' crucifixion. This story is very very weird!

Now let us check the viability of the story of Mary, the Mother of Jesus before we open the book of the Apostle John. Or indeed, we should rather search for details of physician Luke's Jesus first.

Jesus: According to Luke:
Luke 22: 1-2;

> The time was near for the Festival of the Unleavened Bread, which is called the Passover. The chief priest and the teachers of the Law were afraid of the people, and so they were trying to find a way of putting Jesus to death secretly.

Luke 22: 14-23;

> When the hour came, Jesus took his place at the table with the apostles. He said to them, I have wanted so much to eat this Passover meal with you before I suffer! For I tell you, I will never eat it until it is given its full meaning in the Kingdom of Heaven. Then Jesus took a cup, gave thanks to God, and said, take this and share it among yourselves. I tell you that from now on, I will not drink this wine until the Kingdom of God comes. Then

> he took a piece of bread, gave thanks to God, broke it, and gave it to them, saying, This is my body which is given for you. Do this in memory of me. In the same way he gave the cup after the supper, saying, this cup is God's new covenant sealed with my blood, which is poured out for you. But look the one who betrays me is here at the table with me! The Son of Man will die as God has decided, how terrible for that man who betrays him!
>
> Then they began to ask among themselves which one of them it could be who was going to do this.

Luke 22: 54: 'They arrested Jesus and took him away into the house of the High Priest; and Peter followed at a distance.'

Luke 22: 66;

> When day came, the elders, the chief priests, and the teachers of the Law met together and Jesus was brought before the Council.
>
> Tell us, they said. Are you the Messiah? If I tell you yes, you will not believe; and if I ask you a question you will not answer. But from now on the Son of Man will be seated on the right of Almighty God. He answered them; you say that I am.'

Luke 23: 1-5;

> The whole group rose up and took Jesus before Pilate, where they began to accuse him: we caught this man misleading our people, telling them not to pay taxes to the Emperor and claiming that he himself is the Messiah, a king. Pilate asked him are you the king of the Jews? So you say, answered Jesus. Then Pilate said to the chief priests and the crowds, I find no reason to condemn this man. But they insisted even more

strongly, that with his teaching he is starting a riot among people all through Judea. He began in Galilee and now has come here.

Luke 23: 13-17;

Pilate called together the chief priests the leaders and the people, and said to them, you brought this man to me and said that he was misleading the people. Now I have examined him here in your presence, and I have not found him guilty of any of the crimes you accuse him of. Nor did Herod find him guilty, for he sent him back to us. There is nothing this man has done that deserves death, so I will have him whipped and let him go.

Luke 23: 18-24;

The whole crowd cried out. Kill him! Set Barabbas Jesus free for us (Barabbas had been put in prison for a riot that had taken place in the city, and for murder.)
    Pilate wanted to set Jesus free, so he appealed to the crowd again. Crucify him! Pilate said to them the third time, but what crime has he committed? I cannot find he has done anything to deserve death! I will have him whipped and set him free. But they kept on shouting at the top of their voices that Jesus should be crucified, and finally their shouting succeeded; so Pilate passed the sentence on Jesus that they were asking for.

Luke 23: 26, 33, 46;

The soldiers led Jesus away, and as they were going they met a man from Cyrene named Simon who was

> coming into the city from the country.' They seized him, put the cross on him and made him carry it behind Jesus. When they came to the place called 'the Skull', they crucified Jesus there. Jesus cried out in a loud voice; 'Father! In your hands I place my spirit!' He said this and died.

Luke is the only witness who claims that the cross on which Jesus was crucified was carried by someone other than Christ himself. After the above crucifixion drama, the Resurrection drama followed. It should be remembered always that as far as Jesus is concerned the ancient predictions of the coming Messiah, his death, and resurrection *must* be seen to be enacted by him personally and perfectly.

Luke 24: 1-12;

> Very early on Sunday morning, the women went to the tomb carrying the spices they had prepared. They found the stone rolled away from the entrance to the tomb, so they went in, but they did not find the body of the Lord Jesus. They stood there puzzled about this, when suddenly two men in bright shinning clothes stood by them. Full of fear, the women bowed down to the ground, as the men said to them, why are you looking among the dead for one who is alive? He is not here. He has been raised. Remember what he said to you when he was in Galilee. The Son of Man must be handed over to sinful men, be crucified, and three days later rise to life. Then the women remembered his words, returned from the tomb, and told all these things to eleven disciples and all the rest. The women were Mary Magdalene, Joanna and Mary the mother of James; they and the other women with them told these things to the apostles. But the apostles thought

> that what the women said was nonsense, and they did not believe them. But Peter got up and run to the tomb; he bent down and saw the linen wrapping but nothing else. Then he went back home amazed at what had happened.

That the son of man 'Must' be handed over to sinful men, be crucified and three days later rise to life. The author of Luke is demonstrating to his readers that the actor of Jesus character is acting strictly under the directorship of biblical prophesies. He *must* by all means be betrayed, be killed and then rises up again in his original human form; this naturally proves that Jesus was not an ordinary human but the true Son of the Heavenly Father. Note that similar efforts were made by Mary and Joseph to make Jesus's birth as mysterious and relevant as possible. Note again that the three women seen at the tomb site excluded Mary, the mother of Jesus. Mary Magdalene's name consistently appears in all the four Gospels as the only significant female in Jesus' life even immediately after his crucifixion. Again two men in bright shinning clothes appeared to them when the three women were *already* at the tomb site.

Luke 24: 13 says,

> On the same day two of Jesus followers were going to a village named Emmaus, about eleven kilometers from Jerusalem, and they were talking to each other about all the things that had happened. As they talked and discussed, Jesus himself drew near and walk along with them; they saw him but somehow did not recognize him.
>
> Jesus said to them, what are you talking about to each other, as you walk along? They stood still, with sad faces.
>
> One of them named Cleopas, asked him, are you the only visitor in Jerusalem who doesn't know the

> thing that had been happening there these last few days. What things, he asked? The things that happened to Jesus of Nazareth, they answered. This man was a prophet and was considered by God and by all the people to be powerful in everything he said and did. Our chief priest and rulers handed him over to be sentenced to death, and he was crucified. And we had hoped that he would be the one who was going to set Israel free! Besides all that, this is now the third day since it happened. Some of the women of our group surprised us; they went at dawn to the tomb, but could not find the body. They came back saying that they had seen a vision of angels who told them that he is alive.

Verses 26, 27; 'Was it not necessary for the Messiah to suffer these things, and then to enter his glory? And Jesus explained to them what was said about himself in all the scriptures, beginning with the book of Moses and the writings of all the prophets.'

It is important to remind the reader of the one who is narrating this hearsay story. This is Luke who has already warned readers that he is *not* giving an eyewitness account. But after reading through his report, there is every indication (choice of words, sentences cited in quotation, and emphasis on words from Jesus mouth) that he was actually present when these events occurred.

Verse 32, 34, 36-46;

> They said to each other, wasn't it like a fire burning in us when he talked to us on the road and explained the Scriptures to us?' They got up at once and went back to Jerusalem, where they found the eleven disciples gathered together with the others and saying, the Lord had risen indeed. He had appeared to Simon! While the two men were telling them this, suddenly the Lord himself stood among them and said to them, peace be with you. They were terrified, thinking that

they were seeing a ghost. But he said to them. Why are you alarmed? Why are these doubts coming up in your mind? Look at my hands and my feet, and see that is I myself. Feel me and you will know, for a ghost doesn't have flesh and bones as you can see I have. He said this and shows them his hands and his feet. They still could not believe they were so full of joy and wonder; so he asked them, have you anything here to eat? They gave him a piece of cooked fish, which he took and ate in their presence. Then he said to them, these are the very things I told you about while I was still with you; everything written about me in the Law of Moses, the writings of the prophets, and the psalms had to come true. Then he opened their minds to understand the scriptures, and said to them, 'This is what is written: the Messiah must suffer and must rise from death three days later.

Verse 51: 'As he was blessing them, he departed from them and was taken up into heaven.'

Very impressive; First question: if Jesus actually knew his killers and tormentors (the chief priests, elders, Judas Iscariot and Pontius Pilate), and he really wanted to show who he really was, {the real Son of God): what prevented him from appearing to his killers, after all, he could not be killed twice, rather than his close associates who already knew him as the prophesized Messiah? Pontius Pilate, the chief priests, and indeed the entire rabbinical council, would have been the first choice of Jesus appearance after death: but then, of course, his historical and theological role as the Son of God would not have been properly played accordingly.

In summary, the author of Luke can be congratulated for adding more spices into the biblical story; an inclusion which neither add to nor subtract from the basic tenet of the Holy Bible; the amplification of God of Israel's awesome power, the manifestation of human Jesus into the invisible Christ through death, rebirth, and ascension to heaven to live with his father. I have read nothing new yet. Jesus Christ is still a Jew, God of Israel still remains a Jew, God of Abraham, Isaac, and Jacob still the same old Jews, and so was The Genesis God—the inventor of Adam and Eve—also a Jew. Luke himself was a Jew and he has no convincing report in his book that the God in his culture must necessarily be accepted as the God of other cultures as well. I like to repeat that up to the end of the book of Luke, nothing resembling a cosmic god has been detected. For all those who are not convinced yet, that the God of Israel or the biblical god is a god for Israelites only, but rather the God for the whole universe, please read the book carefully. The book itself has said so, in so many chapters and verses. Some of the quotations are already cited in this discourse.

You may encounter such generic words as 'world', 'whole world', 'everyone', 'mankind', 'humanity' and several other such plural generic words in the Holy Bible. Be cautious! The applications of such plural words are common in all languages or dialects. They are always used in reference to the specific cultural frame sense in which the speaker is addressing the issue. Biblical usage of those pluralistic words is meant for the people of Israel or of Jewish ancestry, the Jewish world, and Jewish humanity.

For instance, if a sixty year-old American comments on twenty-first-century teenage fashion, where part of a female buttocks and her underwear are purposely exposed to the public; that, 'what is becoming of this world!?' This may be taken to mean the entire globe, but of course he is making a specific observation in possibly at Time Square area in Manhattan of New York City. The 'world' here is of course used in reference to teenage fashion in midtown Manhattan of New York City environment; definitely not the globe. Teenagers in other cultures may have copied the fashion into their cultures. Bible readers and Christians, specifically, have globalised biblical God stories far beyond human imagination. The book itself is preposterously seen today as a sacred matter or a divine object which is worshiped daily by many Christians, because some believe that every word written in the book is God inspired. There are scores of people who cannot read nor write, but who go to bed with a copy of the Holy Bible tucked under their pillows, because it is believed to contain spiritual protective powers. They refuse to accept the obvious fact that the entire book is like other written works. Again, some believers cannot accept the fact that biblical authors are like news reporters who claim to have head the stories which they write about in the Bible. Every script from Genesis to Revelation is written in report order.

The authors are *reporting* what they have seen, or heard and read about what they write.

The authors themselves, particularly Luke 1, 1-4 has clearly stated that he or she is not an eyewitness to what is written, yet it is believed by many that the Bible is not man-made.

The attachment of the word Holy to Christians' Bible is grossly camouflaged to the truth about the book. It gives instant impression to many that it is a divine entity; a Holy book. Many Bible readers therefore begin to read the book with this preconceived belief and therefore do not read the Holy Bible as an ordinary book, hence their brains do not capture the core meaning of the words, sentences, chapters and verses written in it. I challenge every reader who can cast away the attached holy screen of the Bible to go back to re-read the book with clear conscience; I am certain that the real truth of the so-called Holy Bible will be exposed. It is a man-made object, a human imagination, carefully and purposely written by ancient expert theologians to convince readers of the existence of a single God for all mankind.

iv.) The Book of John (Christ, Word, God)

This is the forth among the four main ancient Gospels selected as the principal witnesses whose testimonies of biblical stories are considered truthful and dependable by theologians. It was written about ninety to one hundred (90-100) years after the death of Jesus Christ.

The author's name is not mentioned. And needless to say, the author *could not claim* to be an eyewitness of whatever he or she is about to tell us, because the actual events occurred about 100 years before this report was written. It must be remembered all along while reading this story that, three other dependent witnesses have narrated their hearings of the same story; this will help to know their similarities and differences. Each of the four Gospels was written in approximately ten years intervals, starting from about fifty years after Jesus' death. Historians of the Holy Bible are in consensus that the book of Mark was the first because some references in Matthew indicate that the author had read the Mark's book. Besides, it is known that book of Matthew did not come out until after sixty years of Jesus's crucifixion.

Again, all of us have heard and used such words as witnesses, and testimonies many times, and by definition these words imply personal or direct knowledge of something. Posterity has nevertheless uncritically accepted these four Gospels even though none claims to have personal knowledge. We must obviously be careful when we cite their quotations.

The Gospel of John presents Jesus as the eternal Word of God, who became a human being and lived among us on planet earth. As the book itself says, this Gospel was written so its readers might believe that Jesus was the promised Savior, the Son of God, and that through their faith in him they might have life.

After an introduction that identifies the eternal Word of God With Jesus, the first part of the Gospel presents various miracles which show that Jesus is the promised Savior, the Son of God. These are followed by discourses that explain what is revealed by the miracles. Mark, Matthew and Luke presented a rapid succession of incidents, of disputes and of shot sharp sayings by Jesus; in John we move to a stately series of set-piece miracles and long, spiraling discussions. Just five obvious miracles are described by John; only one of them, the feeding of five thousands people in the desert is clearly known to the other witnesses. John's Jesus concludes his public teaching with a climatic miracle: he raises from death his friend Lazarus. The crowd is enthusiastic, the Authorities are frightened, and they initiate maneuvers that will lead to Jesus' death. This account is uniquely John's. Sometimes it is doubtful if the four are talking about the same Jesus. Closing chapters tell of Jesus arrest and trial, his crucifixion and resurrection, and his appearances to his disciples after the resurrection.

Let us begin:
According to John 1: 1-5

> Before the World was created, the Word already existed; he was with God, and he was the same as God. From the very beginning the Word was with God. Through him God made all things; not one thing in all creation was made without him. The Word was the Source of life, and this life brought light to mankind. The light shines in the darkness has never put it out.

John 1: 14 *says, 'The Word became a human being and, full of grace and truth, live among us.' Verse 18; 'No one has ever seen God.'* The reader is invited to compare the following.
According to Genesis 1: 1-5

> In the beginning, when God created the universe, the earth was formless and desolate. The raging ocean that covered everything was engulfed in total darkness, and the power of God was moving over the water. Then God commanded, let there be light, and light appeared. God was pleased with what he saw. Then he separated the light from the darkness, and he named the light "Day" and the darkness "Night" Evening passed and morning came, that was the first day.

Genesis 1: 1-5; and John 1: 1-5; are the two most important quotations in the Holy Bible with regards to creation, because they contain the core perception of God and the Universe' creation as understood by ancient Jews or as according to ancient Jewish culture. Although the two author's identities are not clearly exposed, readers are enticed to avoid questioning the contents of the books. Even though the books are man-made and therefore open to human mistakes; besides, the Holy Bible contains thousands of riddles which human brain cannot overlook or simply ignore as irrelevant, or because of religious faith we need not question it.

The author of the book of Genesis claims to know how the universe was created by someone known as God; and that before creation, the earth had no shape and *nothing existed*. The author of the book of John is also telling us that before God invented the universe *something already existed;* and that thing was 'Word'. We are further informed that 'word' is a 'He,' and that he was with God and also the God himself. Of course in fictions characters can be given any name as long as their expected roles are performed accordingly.

The Holy Bible is by any definition, a book of fictions based on man-made stories or a human being's imagination of how the universe came about. Genesis' author thinks the universe was invented by a character called God, who is the ultimate inventor of all things in this world. It is a he God who has a perpetual existence and also operates through human beings and Angels. This God exists only because the writer of the book of Genesis says so. He has no supporting evidence to substantiate the claim.

We really do not need to be told that John's author has created a fiction. A fictitious character which he or she, the author, has personally created: again we must bear in mind that this author is not saying that this idea came to him or her through dreams, nor a message delivered from God through an angle. He is also not reporting to us from what he has read in the past. He is therefore the Author or the Originator of the '*word*' concept. Unfortunately we are not informed who this fictional character writer was; hence it is up to you and me to assess this ancient fantasy. In the past, some theologians claim that book of Genesis was written by Moses, while the book of John had Paul as its author. If this is proven true, our argument regarding Bible as a fiction would simply be irrelevant, because we would know and accept the Bible as an obvious human product.

This of course is definitely not the first time the Holy Bible has come out with a mysterious story, so we should not be very surprised. But let us examine the logic behind this fiction: if the term 'Word', 'He', or John's 'God' existed before the universe was created, and if so, do we need to be told this? Genesis has already invented one for us; or, are we to understand that John's God was there before Genesis' God? Were there two Gods and is that the reason why the Genesis God refers to '*now we will make human beings, they will look like us and resemble us*'? Do the 'we' and the 'us' *refer to them both?*

Les us read John 1, 14; '*the Word became a Human Being and full of grace and truth, lived among us. We saw his glory, the glory which he received as the Father's only son*'

The author is very explicit with the meaning given to the character Word; it is definitely referred to Jesus, *the Father's only son*. If Jesus was in the universe before God created the universe; the *human being* Jesus *who lived among us,* who had died about one hundred years prior to writing the book of John, was actually alive millions of years when the so-called creation took place: the author is either out of his mind or the logic in his fiction is nonsensical. John's idea of Jesus is nothing other than a fictitious character which was never born, never lived among us, never died, never resurrected, and never ascended to Heaven as other witnesses have already misled Bible readers.

This indeed confirms the Gnostic scriptures of Mary's virgin birth as unrealistic. One of the two midwives was attempting to physically examine Mary after the *moment of delivery* if Mary was indeed still a virgin. If indeed Mary was still a virgin and baby

Jesus was there with Mary, Joseph and the other midwife, from where baby Jesus appeared this world! The whole Jesus episode plus God's creation as already analyzed in this discourse are human imaginations. I have already stressed that the content of the Holy Bible is essentially sectarian. It tells us how ancient Jews understood their world. The book does not and cannot explain how other cultures understood their world. It will be naïve for any non-Jew to assimilate biblical concept as a universal precept. The universe which the authors refer to is the universe of the twelve tribes of Israel. It cannot be applicable to all cultures of the world. Absolutely impossible!

Acts 7: 8; *'God gave Abraham the ceremony of circumcision as a sign of the covenant. Abraham circumcised Isaac a week after he was born: Isaac circumcised his son Jacob, and Jacob circumcised his twelve sons, the famous ancestors of our race.'* The Bible contains biblical God in Genesis to Adam and Eve, to Abraham, Isaac, and Jacob, the twelve ancestors of Jewish race, their traditions and customs nurtured over the years. It continued to the conception, birth, works, death, and resurrection of Jesus Christ who later became the son of god and God himself. This is all what the Bible is about. The universe it talks about is the cultural world of Jews. It is definitely not the global universe, the total cosmic planet-earth in which you and I and Jews, separately, are a mere infinitesimal part. This is not a God for the Chinese, the Indians, the Africans, the Eskimos, the Japanese or indeed any other planetary person.

I like to remind readers once again about the original aim of this discussion. It is solely an attempt to establish the plausibility of the existence of one God, a single God who is supposed to have created the globe; particularly the biblical God. John's author is obviously a mystique writer with plenty of fantastic stories. We have already discussed a few, but now lets go to Christ' narrative as illustrated in the Gospel According to John. This is to find out if there is a reasonable ground to believe what is written in the book.

Over a thousand years after the death of Jesus Christ; particularly since the introduction of Apostle John's book, that is, about a 100 years after Jesus' death, the word or the concept known as 'Christ' had been hotly debated endlessly by several intellectuals at the time and there was no consensus of a single definition for Christ as a name or a concept.

The word has become a specialized field of study in Christian theology and it is classically termed 'Christology'. *To be discussed later.* In the mean while let us examine Mary, the mother of Jesus or possibly the mother of God.

Chapter 7

Mary

i.) Mother of Jesus Or Mother of God?

Let us turn away a little from the Holy Bible which contains only synoptic scriptures, and switch to the Gnostic Bible to find out what the Gospel of Mary (Jesus' mother) has to say about the birth of Jesus. We need to do this because Holy Bible deliberately omitted many pertinent stories about significant biblical people. 'This document is the source of almost all the details we believe we know about the life of Mary, the mother of Jesus Christ. It also sets forth an account of the birth of Jesus that is much different from the Christmas story accepted around the world today. Written around 150 AD by an unknown Christian; it was widely known in many parts of the early Christian world; yet few people today have ever heard of it:' A quotation from the Book.

While the New Testament focuses on Mary and Joseph, in the context of the birth of Jesus, the Infancy Gospel of James begins years earlier by describing the plight of an elderly couple named Joachim and Anne. Though prosperous and prominent, they were childless, a condition not acceptable to the society of their time. In fact, Joachim alone of all the good people in Israel had no children. After Joachim is publicly chastised for his lack of offspring and his offerings to God are rejected by the priests, he decides to retreat into the wilderness to fast and to pray in the hope that God will tell him why he is childless. His wife Anne is similarly blamed by her own slave for being barren. Stricken with sorrow, she sits in her garden only to be further reminded of her barrenness by the fruitfulness of everything around her.

An angel hears Anne's lament and informs her that she will bear a child who will be known throughout the world. In the desert, Joachim is likewise told that his prayer has been heard. Overcome with joy, Anne vows to dedicate her child to God, and Joachim rushes from the desert to be united with his wife. The child of course is Mary, who became the mother of Jesus.

After Mary's birth, Anne transforms the baby's bedroom into a sanctuary where nothing unclean can touch her. On the young girl's first birthday, the priests blessed her. Then when she was three, her parents fulfilled their vow by presenting Mary to the priests in the Temple in Jerusalem. There she spent the rest of her childhood, and needless to say, none of this appears in the New Testament. Even when the Infancy Gospel begins to overlap with material in the Gospel of Matthew and Luke (the only books in the Holy Bible that describe the circumstances surrounding Jesus' birth), it tells a very different and much more detailed story. At age twelve, Mary is about to become a woman (*beginning of menstruation*) and hence a threat to the Temple's purity. Having sought divine guidance, the high priest is instructed to summon all the widowers of Israel, each to bring his staff. A sign will determine which of the suitors is to receive her as his wife. Among the widowers was Joseph who in this account is described as an old man with grown sons.

A dove appears at the tip of Joseph's staff and perches on his head, indicating that he is the one chosen to become Mary's husband. Joseph protests that he is elderly, with grown children, and does not wish to remarry. The priests agreed to let him take Mary as his ward, not as his wife. Joseph took Mary home at age twelve. Joseph's job as a carpenter or a builder calls for him to go out of town to build houses.

ii.) Mary: Mother of Jesus is pregnant

Together with other virgins, Mary is asked by the high priest to spin thread for a new Temple veil. While drawing water from a well, Mary hears a voice, terrified, she hurries home to her spinning. There the voice, that of an angel, addressed her again, saying that she has favour with God and will conceive by means of his word. From here on, the story is the same as Luke's version in the New Testament; where Mary visits Elizabeth, Elizabeth' pregnancy, to the birth of John the Baptist. Another diversion begins when Joseph found out Mary was pregnant. Initially, he blamed himself for failing to protect the girl. Mary declared her innocence but Joseph would not accept her until he himself had a dream in which an angel tells him that Mary is pregnant by the Holy Spirit and that he is to name him Jesus. Obedient to the Angel's message, Joseph abandons his thoughts of divorce.

A major crisis soon develops when a visitor realizes that Mary was pregnant and tells the high priest that Joseph has violated the virgin in his care. Joseph and Mary were summoned to the high priest but both denied. The priest refused to accept whatever they told him. Joseph was asked to return Mary back to the Temple. Joseph wept. Up to this point Matthew, Luke, and James seem alike; many significant departures set into the story when Joseph and Mary heed Emperor Augustus' call for a census and head for Bethlehem, same as in Luke. There is a conversation between Mary and Joseph while they travel. Most importantly they stopped *before* they reached Bethlehem so that Mary can deliver her child. Not in a stable *but in a cave*. Also included is a miraculous vision which Joseph experienced at the moment of Jesus' birth in which 'time stood still'. Clouds and birds are frozen in the air, and men and animals remained motionless on the ground. Where Luke tells of shepherds arriving from their fields, the Infancy Gospel of James has two midwifes who visit the cave where Mary has *already given birth*. Skeptical of the claim of one midwife that a virgin has given birth, the second midwife (Salome) performs a physical examination by inserting her finger into Mary to confirm that the child has indeed been born to a virgin. But as she touches Mary, her hands begin to burn, and she prays for help. A voice tells her to pick up the baby Jesus. When she did, her hand was healed.

As Joseph, Mary, and the infant Jesus are preparing to *leave for Bethlehem after delivery*, the Magi arrived at the cave in search of the new born king of Judeans.

According to Gnostic scriptures: two things either happened or did not happen; firstly Jesus was not born in Bethlehem as prophesizes in synoptic scriptures; secondly, that according to the two midwives who did not assist, but were at the cave where Mary delivered baby Jesus, Mary was still a virgin even after delivery; question is, from where did the baby enter the world? The answer lies in the grave yards of first. and second Century's religious fanatics; they carefully crafted Jesus Christ as the visible likeness of the invisible God. In this sense neither Jesus nor the biblical Christ ever really existed. At least Jesus was not born in Bethlehem; a vital disqualification of the so-called Isaiah's prophesied messiah.

Jesus has suddenly become adult; a rabbi, a miracle performer and a prophet who speaks with godly powers. He had made for himself a few followers, and several enemies among the Israelites; the Jewish priests who saw Jesus as an imposter and to the Emperor's provincial King Herod and Governor of Galilee Pontius Pilate who are informed that Jesus was provocateur. His days in the visible world became numbered:

Now there is a plot to kill Jesus and it is revealed by Jesus himself; because most of his teaching was focused on anti-Jewish traditions and also not in line with laws of Moses.

CHAPTER 8

Jesus of Nazareth

Who or What was He?

'Mary's Son; God's Begotten Son; Christ; Messiah;, Anointed One; Rabbi; 'Word'; Holy Ghost; Lord; God; Magician; Husband; Father?'
Was Jesus one of the above, some of the above, or both at the same time?

There is a consensus among all the New Testament Gospels that Jesus was at least ten of the above virtues. There is a general disagreement of the last three question marks; a magicians a husband and a father. Let us thenceforth focus on the three remaining Jesus' unmentionable merits.

There is a record that in the Bible and elsewhere that several religious leaders in Palestine province of the Roman Empire performed miracles in public places. The narrated Jesus' miracles in the Bible were not unique. Popular miracle performers such as Simon and Mithras had followers who revered them as Gods or Sons of God. Biblical writers preferred to call the illusions created by religious leaders as miracles rather than magic to mystify Jesus Christ's image. It is not demeaning to classify Jesus as a magician. Jesus characterisation above as *a Magician* is quite befitting.

With regards to Jesus as *a Husband* and *a Father*, both Synoptic and Gnostic Bibles contain several clues which make such characterisation fit the biblical human Jesus Christ. Let us now examine the evidence.

The story begins from the book of John, the fourth Book of the New Testament: There appear to be an unholy passionate feud between the immaculate Jesus and his mother Mary and the other Mary in a Wedding Party.

According to John 2, 1-11

> Two days later there was a wedding in the town of Cana in Galilee, Jesus' mother was there, and Jesus and his disciples had also been invited to the wedding. When the wine had given out, Jesus mother said to him, they have no wine left. You must not tell me what to do, Jesus replied. My time has not yet come. Jesus' mother then told the servants, do whatever he tells you. The Jews have rules about ritual washing, and for this purpose six stone water jars were there, each one large enough to hold a hundred liters. Jesus said to the servants, fill these jars with water.
>
> They filled them to the brim, and then he told them; now draw some water out and take it to the man in charge of the feast. They took him the water, which now had turned into wine, and he tasted it. He did not know where this wine had come from (but of course the servants who had drawn out the water knew); so he called the bridegroom and said to him, everyone else serves the best wine first, and after the guests have had plenty to drink, he serves the ordinary wine. But you have kept the best wine until now.

Please my dear reader, try and re-read the above quotation again; this time however, very carefully: watch the sentences, the chain of thought, the type of questions asked and who ask them;

their answers, the invited guests, the *others who were already there*; the bride and the bridegroom, and the overall significance or structure of the texts. Your general quick conclusion may be that Jesus went to a wedding party in Cana Galilee; or as generally concluded by Bible readers, that Jesus performed his first miracle by changing water into wine.

Beyond this first impression, you also notice that the author is very meticulous about the choice of words, construction of sentences and presentation of the story. It appears John's author intends to talk about the first miracle Jesus performed: but really, John need not tell us where and how it took place, as indeed several examples of such miracles are told without detailed account; for instance *John 6, 16-21,* Jesus walks on water and very little information is given. Again in *John 9, 1-7* Jesus heals a man born blind, and very few information is exposed to readers. In this particular miracle where water is turned into wine, there appears to be a need to inform readers of all the circumstances leading to the miracle. At the same time, there appear to be a feeling on behalf of the author that too much information is becoming exposed.

A wedding took place in a town called Cana in Galilee; but who's wedding? The author appears to know the names of the bride and the bridegroom but readers are not told who they are. Why are the names concealed? The reason is not likely to be lack of knowledge of them, but rather a calculated intent to withhold the names from public knowledge.

The importance of public interest regarding the wedding couples' identities stems from the already mentioned sacred names as Mary the mother of Jesus who we are told *was there* present at the wedding party. Note that she was not among the invited guests. She was just there; but in what capacity was she there? The author is not saying that she lived there or that she was the host who organized the wedding party, and if she did, on whose behalf was it done, her daughter or her son Jesus? It is most likely to be her son because none of the Gospels ever mentioned a single name of Jesus' sisters or Mary's daughters; although a certain Mary is mentioned and implicitly, she is Jesus' mother [Mary's] daughter. Again readers of the Bible are *not* told that *Jesus ever* married and neither are we told that *Jesus never* married. One other interesting part of this is the invitation to Jesus' disciples to the wedding party. This implies that *both males and female disciples attended the wedding party*; including the famous companion of Jesus, Mary Magdalene. Again the author of John is not explicitly saying the female disciples were excluded. But let us assume that the females were not invited.

It is mentioned elsewhere in this work that Jewish traditions and customary rights permits no male to be honored the title of a Rabbi if he is not married and has no children. This tradition was there before Jesus was born and it is still observed in Jewish communities all over the world even today. Could Jesus be deemed unholy or a sinner if we were told he ever married a woman? What is so sinful about that? Who knows, perhaps [His] celestial holiness would be tarnished if he did.

The Mother of Jesus is informed by servants of the wedding party that wine is given out or that there is no wine left to serve. Here again there appears to be a universal convention that in all parties or ceremonies of this nature it is the host who provides logistics such as food, drinks, chairs, tables, music and the rest for the enjoyment of everyone present. It is definitely not the duty or a requirement of anyone present, especially invited guests, to see that shortages are replenished inconspicuously. It is strongly suspected that the servers told Jesus's mother because she was the bridegroom's mother who conducted or organized the wedding for her famous son [Jesus]. Notice Jesus's reaction when his mother Mary told him that there was no more wine left; 'you must not tell me what to do! My time has not yet come.' Jesus is angry with his mother for no apparent reason. Jesus response to his mother's statement is completely out of context. It is also fascinating that this is the *only time in the entire Bible* that Jesus (as an adult) and his mother ever communicated verbally.

What is the relationship between Jesus' statement 'my time has not yet come' and 'there is no more wine left' statement from his mother. Mary's statement does not request miracle performance directly or indirectly from her son Jesus. Besides we are told Jesus was an invited guest anyway, and both he and his mother had no business interfering with the wedding event in anyway. It is not stated anywhere in the text that the host or anyone present begged Jesus to perform a miracle or make wine available to the guests.

If Mary was not the mother of the bridegroom and Jesus was not the bridegroom, the wine shortage information could have been addressed to *the man in charge of the feast*. According to the Gospel, someone special had been assigned to play the master of ceremony's role; or surely the parents of the bride and the bridegroom would have been informed about the wine shortage.

On top of all that, the Mother of Jesus told the servants to do whatever Jesus tells them to do.

An invited guest 'Jesus' and the guest mother Mary could not give command to servers in a wedding party organized by the bride and the bridegroom parents. Practical common sense shows that the wedding party belonged to no one other than Jesus and his mother and of course the bride and her family.

Finally when Jesus told the servants to take some of the water to the man in charge of the wedding feast, there it was realized that the water had turned into wine. And after tasting *it the man in charge of the feast told Jesus* that most people on such occasion do serve the best wine first and when everyone has had enough to drink he serves the ordinary drink.

This last conversation was between Jesus and the master of the wedding ceremony; *in addressing Jesus* he said '[you] have kept the best wine until now' This clearly shows without doubt that Jesus was the bridegroom, otherwise why must an invited guest like Jesus and nobody else keep and then release or control drinks at someone else party.

We must not forget that the name of the bride is also missing from the New Testament. Holy Bible readers are definitely entitled to know the name of the woman who was getting married. It is a serious insult to readers' intelligence to speak of a wedding feast without mentioning the names of the married couple. In *Matthew 22, 1-14*; Jesus talks about *his own wedding feast with invitation to everyone* in parables so that, as in *Luke 8, 9-10*, we shall not understand the true motive of what he says in parables; but this wedding at Cana as narrated in the book of John is the one which ought to have mention Jesus as the bridegroom: rather Jesus is treated as a mere wonder guest who shows off his magical powers in presence of dignitaries: turning water into wine as his first ever miracle. Indeed Jesus probably deliberately chose his wedding day to impress his wife and family members plus the invited dignitaries with his so-called God given powers.

Who the bride is remains a mystery in the Bible too; but there are two possible suspects we shall look into.

The wedding was indeed a big one and by the author's narratives, a lot of very important dignitaries must have been invited considering the status of Jesus as a very knowledgeable personality, a rabbi and a descendent of King David. Above all, he is acclaimed to have been born by virgin birth. The house in which the wedding took place for instance possessed large stone jars. At that time, that is about 2,041 years ago, any house in that Province of the Roman Empire that has six large stone water jars would be a mighty mansion. The family that owns such property must be considered exclusively affluent. The one person among all Jesus followers that all the Apostles in the New Testament testify as the financier of Jesus and his entourage and also Jesus' closest companion was Mary.

Six large stone jars each containing 100 liters (about a minimum of one hundred bottles each); when multiplied by six, it is 600 liters or at least 600 bottles of wine. In modern standards, a six hundred bottles of wine party would certainly be classified the most sensational wedding of the century. It is no surprise the author considers it necessary to give readers a hint of the famous wedding which took place in a town called Cana, rather than just Jesus' miracle of changing water into wine alone. The miracle story alone would have been enough if the wedding was insignificant.

Now, so far so good, for the bridegroom: but who might be our first bridal suspect; the woman who could possibly be considered as the wife of Jesus or Mrs. Jesus Christ sounds pretty odd and funny. We shall nevertheless glance through the Synoptic Gospels first and then go to rest with the Gnostic Bible.

Synoptic scriptures throw a sharp clear bright light onto two possible biblical women who fit the description of the image of biblical Jesus Christ' wife; the first one is Mary of Magdala.

Those of you, who have *read* the Holy Bible, book by book, from Genesis to Revelation but not just chapters and verses, will not be surprised of this choice. The second possible female will be Mary of Bethany, the sister of Martha and brother of Lazarus.

Both women are from different towns, Cana and Magdala, in Galilee province and they are also descendants of Benjamin tribe.

The relationship between Jesus and Mary of Magdala on one side, and his relationship with Mary of Cana or of Bethany on another side are found in several disjointed chapters and verses in the New Testament. Jesus' links with these two women have many significant biblical implications that ancient writers of Holy Bible preferred not to expose to the reading public for fear of tarnishing the saintly gracious painted image of Jesus or the Christian faith.

The claim that Jesus Christ had a secret wife or wives used to be a heretical offence. It was punishable by death during the early Middle Ages. It was a serious offense because the authors of the New Testament have written that Jesus was not just Christ but the only begotten son of God. And as such has no need for a wife. I do not see much reason why this is necessarily a blasphemy, because the Holy Bible and the Gnostic Bible have given many clues to this effect that the human Jesus had a romantic relationship with a woman or indeed women during his lifetime on planet Earth. Why must a real human being with the same male hormones and male sexual organs like any planetary human male necessarily be portrayed as sexually inactive? Should we accept the celibacy characterisation of Jesus as parts of biblical fiction as narrated by the Gospels writers or should we assign Jesus some real human characterisation that resemble you and I? Besides, there is nowhere in the Bible where it is written directly or indirectly that Jesus never or ever married; or never or ever fathered a child or children. These same Gospels have shown Jesus' advocacies of marriage as a blessed human institution worthy to mankind. Why then would he in person not be married if his supposed heavenly Father approves it? This makes no sense. The same senselessness pops out when we look for the whereabouts of Jesus in the Bible between age thirteen and twenty-nine; was he in hibernation with his wife and children somewhere in Bethany? But why not!

According to Hebrew scholars a teacher is a Rabbi. Jesus was a teacher, hence Christ was a Rabbi. And according to Jewish Mishnaic Law and customs, an unmarried or a childless man cannot be a Rabbi, hence as a Rabbi Jesus Christ actually did marry and had children being honored as a Rabbi. The Holy Bible never said Jesus Christ never had a wife or children. If this was so the Bible would have said so: the Bible would also have mentioned if he had them.

Matthew and Mark tried to avoid mentioning Mary Magdalene's name as if she was insignificant in Jesus' life by merely stating that she accompanied Jesus during his teaching trips to Judea and elsewhere, and her presence at the crucifixion and resurrection sites.

In Luke, it is written as if Mary Magdalene was a devil possessed with the so-called seven demons which Jesus exorcised.

Some revised modern Holy Bibles state that Mary Magdalene was a prostitute. In fact the Catholic Church waited until 1969 before declaring publicly that Lady Magdalene was not a sinner or a prostitute. All these years, this Mary's image had been stained with filth and mud simply because a few top rabbinical leaders need to fulfill celibacy agenda for their human God.

But the same Gospel of Luke, in its earlier chapter describes 'a woman' who anointed Jesus with very expensive perfume and used her long hair to clean Jesus feet. In Jewish culture, women do not anoint men, hence even if this event actually happened; it should not be made public. Gospel of Mark also made mention of 'a woman' who anointed Jesus with spikenard, an extremely expensive perfume the cost of which could feed many poor for a long time as a waste on Jesus' body. Jesus immediately responded: *Mark, 14, 6; 'leave her alone! Why are you bothering her?' 'She has done a fine and beautiful thing for me' Matthew 26, 6-13;* Repeats the same reaction of Jesus; and even though none of the two Gospels mentioned the name of the woman who anointed Jesus, Mary Magdalene and Jesus behaviour regarding Anointment of a Rabbi by a female companion with spikenard, the most expensive Indian perfume, and to use her long hair to clean Jesus toes in Simon's house in Bethany definitely shows a romantic and sexual relationship among the two. In John Chapter 11, Jesus' overall relationship with Mary Magdalene and her family vividly comes out. Jesus Christ' deep affection and inner passion for Mary and her family in Bethany was not an ordinary one. It is also indicative that this is where the *shortest verse* in all the books of the Holy Bible is found *John 11, 35 'Jesus wept.'* The house of Bethany's apostolic romance certainly resonates to the contents of the book of 'the Song of Songs.' The passionate exchange of love talks between a man and a [woman with long beautiful hair which falls along the neck like jewels]; and [the kind sprinkled with my spikenard perfume and lays upon my breast is my darling among all women]; (the First Song). Do we say these songs have nothing to do with Jesus and Mary Magdalene? If the *impressions* given by the Gospels that Jesus was a real human being like all of us, (flesh, bone and blood but not a divine entity) are considered, his human sexuality cannot be taken away from him. With his portrayed biblical image, I would not be surprised if new written documents like the 1945 Nag Hammadi scrolls come out to reveal the complete sex life of Jesus.

Apart from general consensus among the Gospel writers that Mary Magdalene always accompanied Jesus in all his trips, it is written that she was among the very few people present during Jesus' earthly ordeals such as his supposed crucifixion. She was the first woman among the few to visit Jesus tomb site, and the first woman or person whom Jesus Christ revealed himself to after his resurrection. This is certainly not a mere companionship between two consenting male and female adults. There was surely an intimate love affair between them. As already informed, all the above references about Jesus sexual affairs are based on the Holy Bible.

We now switch to the Gnostic Bible to see what is written there; if Jesus Christ indeed had a spouse during his lifetime. Be reminded also that the ultimate purpose of this discourse is an attempt to find out if there is a possibility of a single creator of this universe 'planet earth' and if there is, could it possibly be the biblical God whose likeness is Jesus Christ as portrayed in Christian literature. As far as to this point, the only God discovered is still the Jewish cultural God who has been deliberately transformed or redesigned by ancient theologians into several types of incredible images which are very comparable to other Gods in other cultures. Jesus was as a pagan and a follower of polytheist religion (Jewish cultural religion) just as all of us following our individual cultural religions with our cultural Gods. Jesus Christ' holiness and saintly image and all the so-called miracles were added to his personality after he had died, by Evangelists, decades later after his death. Jesus was no different compared to modern-day religious fanatics operating in the various cultures around the world. They cherish procreation through marriages. Jesus sexuality is deliberately omitted from the Holy Bible to make him resemble the God of

Israel and the Genesis fictitious God. Both of whom are narrated with no feminine relationship.

His lack of sexuality would not necessarily qualify this biblical character, Jesus, to be accepted as the cosmic God. It actually enhances readers' suspicion of the Gospels malicious intent to cover up the true identity of the human Jesus.

CHAPTER 9

Christology and Mariology

Christology is mainly concerned with spiritual personality of Jesus Christ as recorded in Christian literature. It is *not* concerned with Jesus' *humanity biography*; it is concerned with his divine makeup which includes his works, his sayings, his miracles, his death, his resurrection, and his relationship with God—his Father—as narrated in the New Testament. Because of its wide conceptual coverage, and multiplicity of evangelism, it is probably better to look at the 'Christ' idea in three separate episodes for clearer magnification. Greek word for Messiah is Christ or Christos. In both Greek and Hebrew the word Messiah means the anointed one and it was generally used in reference to a King.

Historically, at the time when Jesus was alive the term Messiah and the Anointed one had nothing connected with divine. They were essentially a complimentary or an honorary titles popularly used when addressing a political leader such as the emperor, appointed governors, or a king as in King David. It was never a divine title. This explains why such words as Christ and the Anointed one were never used in the Old Testament's characterisation of the forthcoming Messiah. It also explains why Jesus could not call himself the Christ or the Anointed One or accepted outright whenever he was addressed with either of the two words. He is quoted by the Apostles to have replied to such questions as 'are you the Christ?', 'are you the anointed one?', or 'are you the king of the Jews? His answer was always if you say I am. It was the later years Evangelist particularly Paul who initiated a vision of a Messiah or Anointed one with *divinity* when Jesus had died decades prior to this.

First, the apostolic era Christology; that is, from Christ birth to one hundred years after his death. This embraces the perceptions of Paul, Mark, Matthew, Luke, and John of 'the character Jesus' as seen by them individually. The New Testament shows their differences and similarities

Second, the Post-Apostolic debates which took place between the second and the ninth centuries. This will cover the founding fathers of Christianity debates on the meanings given to Christ according to the four main written languages at the time; *logia* and such words as *Kyrios* Greek for Lord, and *Mari* an Aramaic for Lord, plus *Logos* translations in both Greek and Hebrew: the above technicalities gave birth to another pertinent Greek word *Theotokos'* which translates as the mother of God—Mary. studying the mother of Lord Jesus Christ, it becomes the study of the mother of God; hence 'Mariology' as a concept became very problematic in understanding the relationship between the earthly Jesus, the divine Christ (mother of God) and the Heavenly Father—God.

The third episode is centered on theologians of the Middle Ages (eleventh century to seventeenth century), and possibly, up to the modern-day understanding and application of ancient theological surroundings of Christianity; a) the human being, the earthly creature who became the God of the universe, and b) the heavenly God (word) who became an earthly human being with human flesh as well as spiritual after his earthly life.

Information in the New Testament of the Holy Bible is technically regarded as Synoptic scriptures because it contains specially selected part of the story of Jesus the Christ. The authors of the seventeen books of the New Testament inform readers that they write about Jesus' works, his teachings, his miracles, his sayings, his death and resurrection. Their concern is about incarnation of the man called Jesus. New Testament is not concerned with the total personality or the total biography of Jesus and the rest; for this reason all information about the man regarding where, what or who Jesus was at the ages 'between' one to eleven, thirteen to twenty-nine, thirty-one and thirty-two *are not written in the Bible*. The missing information during these periods about Jesus was considered irrelevant by the early founders of the Christian faith. Nevertheless, a portion of the missing part, over forty ancient writings, was discovered in a town called Nag Hammadi in Egypt in 1945. The stories revealed in this portion of Jesus story fits perfectly into the missing spaces of the Bible.

Thus during the first 150 years of Jesus Christ era several important questions came up among elders of his followers.

The repetitive and ticklish nature of such quotations in all the four Evangelists (Matthew, Mark, Luke and John) 'but who do you say that I am?' With Peter's answer 'You are the Christ,' opens several doors to search for the missing parts of Jesus Christ information not recorded in Holy Bible. Similarly, the forth Evangelist, John, deliberately ignites every reader's mind with a matchstick in sparkling flames with two crackling narratives. John 1, 1-14, as already quoted above is dedicated to the 'divinity' of Jesus with the most wacky word 'logos' which is translated as 'word' with its transformation into a divine entity synonymous with Christ, and most weird of all, with 'word's' cosmic preexistence in the same category as Genesis' God the creator. The other crazy statement by John is the book's last chapter and its last verse, John 21, 25. That the *whole world* would not have enough room for the number books about Jesus Christ if everything was to be written down.

The first three Gospels are however telling us that Jesus was an extraordinary human being because he was conceived by the Holy Spirit in his mother Mary's womb and his death was the death of only his human flesh. That the holy spirit in him went back to where it came from after his death; again, before the spirit' ascension to its source he regained the human flesh back to its full nature as a real person who actually interacted with his lifetime disciples, and with whom he shared meals and conversations together. These narratives make the guy a real superman. This fantastic story really sets in motion millions of questions. The complexity of the above is what brings about Christology as a concept worthy to look into. We are told that Christ is the second name of Jesus; Word is the same as God; Jesus Christ was therefore the God who created himself and all mankind in his own image. What a contradiction!

The second century theologians such as Ignatius of Antioch and Irenaeus translated the entire early evangelical ambiguous words which were originally Aramaic, Koine Greek, and Mishnaic Hebrew into their respective language interpretations and understandings or manipulations to suite their religious agendas. During this period several of the Bishops or religious leaders claimed specialty in their linguist background resulting in the emergence of group formation which depended upon language spoken in the various church districts and selective Evangelical fellowships. Church of Matthew, Church of Mark, Church of Luke, Church of Apostle John, Church of Evangelist Paul and several others with different outlook of Jesus Christ developed.

A number of diverse opinions developed. Opposing groups of Christ followers debated words translation as applicable to each domain. Churches were formed according to which of the four Evangelist' narratives of Jesus is preferred. Struggle to win followers to spread the specific apostle's gospel brought about common feud among early Christians of the second Century AD.

When indeed the entire biblical stories are critically scrutinized, or looked into with magnifying lenses, it becomes very doubtful if any objective reader with unbiased conscience can really rationally associate reverence to the Holy Bible?

Chapter 10

Mary Magdalene and Jesus Christ

i.) What if they were married?

According to Gnostic Scriptures;
The Gospel of Mary Magdalene is a book which has been available in most reputable bookstores since its discovery in Egypt in 1896. Although some portions of the *papyrus* on which the text is written were damaged and partly missing, the remaining readable parts contains vital information about her which are not available in the popular Holy Bible. During Jesus lifetime most people had only one name as compared to modern-day name structures, when most worldly people have at least two names; the first name and the last names.

All the people or disciples who were closely associated with Jesus had singular names, except that almost all of them had some kind of adjectives attached to their names to clarify their identities. Mary's adjectival name was 'Magdalene' in Hebrew it is 'Migdal' which means a 'tower'; and in Aramaic 'Magdala' which is a name of a fishing town. Mary of Magdala or of Migdal over the years became Mary Magdalene. Today, Magdala is a name of a village in central Ethiopia according to The Oxford History of the Biblical World.

It is noted in Gnostic scriptures that all the male disciples who accompanied Jesus were married-men whose spouses were probably always part of Jesus entourage. In Luke chapter 8, we are told there were *several women* who accompanied Jesus in his travels through towns and villages to preach the Good News. Due to Jewish traditions, women's public identities are insignificant and they are always concealed. For instance in 'Templar Revelation' *(refer to bibliography)*, it is written that women's testimony was not allowed in Jewish Courts of law during Jesus era, hence women's word about anything was not considered important.

Mary Magdalene's role among the group was such that the four Gospels would have preferred not to mention but due to her theological and social significance and possibly, because of her personal link with Jesus, she could not be omitted out completely.

Note that none of the so-called *'many other women'* as quoted above who accompanied Jesus have their names and roles concealed by Luke.

There were therefore many other things about Mary Magdalene which are not popularly publicized; especially her ideas about the nature of the universe, the teachings of Jesus, her role as the group financier, as Jesus personal confidant, her relationship with other disciples, her personal life, her age and siblings, her marital status, her family and children, and whatever happened to her after her encounter with the resurrected Jesus. For some undeclared but highly suspicious reasons, Mary Magdalene is the only woman who has been painted with the most bizarre look of a questionable reputation.

ii.) Was Mary Magdalene *Mrs. Jesus Christ?*

Would it be a sin *if she was?*

1 Corinthians 7: 3 states the following: *'There is no sin in getting married.'* There are several books in the Bible that encourage marriages between male and female lovers—Genesis, Ruth, Timothy, Psalms, Song of Solomon, Hebrews, John, Corinthians, and others evidently endorse matrimony as holy. Why can't the supposed 'holiest' of men be a husband or a father? I just don't get it!

A very interesting conversation between Mary Magdalene and some of Jesus' disciples is reported in the Gospel of Mary of Magdala in the Gnostic Bible.

> Peter said to Mary; sister we know that the Savior loved you more than other women. Tell us the words of the Savior that you remember which you know and we do not. We have not heard them.' Mary answered saying; what is hidden from you, I will reveal to you. She began to speak these words, saying, I saw the Lord in a vision and I said to him, Lord, I saw you today in a vision. He answered and said to me, blessings on you, since you did not waver at the sight of me. Where the mind is, there is the treasure. I said to him, how does a person see a vision, through the soul or through the spirit? The Savior answered saying; a person sees neither through the soul nor the spirit. The mind which lives between the two sees the vision.

On another occasion,

> Peter also opposed this. He asked the others about the Savior, did he really speak to a woman secretly without our knowledge, and not openly? Are we to turn and all listening to her? Did he prefer her to us? Levis said to Peter, you are always angry. Now I see you contending against this woman Mary as if against an adversary. If the Savior made her worthy, who are

you to reject her? Sure the savior knows her well. That is why he loves her more than us.

The Gospel of Philip has the following to say about this odd couple:

'The companion is Mary of Magdala. Jesus loved her more than his students. He kissed her often on her face, more than all his students, and they said why you love her more than us? The Savior answered, saying to them, why do I not love you like her?' 'Great is the mystery of marriage! Without it, the world would not be. The existence of the world depends on marriage. Think of sex. It possesses deep powers, though its image is filthy.'

'There is the earthly son, and there is the son of the earthly son. The lord [Jesus] is the earthly son, and the son of the earthly son is he who is created through the earthly son.' 'No one can know when the husband and the wife have sex except those two. Marriage in the world is a mystery for those who are married.' 'If a marriage is opened to the public, it has become prostitution, and the bride plays the harlot. Bridegrooms and brides belong to the bridal chamber.'

Again from another book, Philip said,

And the companion of the Savior is Mary Magdalene. Christ loved her more than all the disciples and used to kiss her often on her mouth. The rest of the disciples were offended by it and expressed disapproval. They said to him, why do you love her more than all of us?

The above narratives or quotations come from a Bible, the Gnostic Bible. The authors are not the real persons who actually wrote them. In other words, both the Book of Philip and the Book of Peter were not written by Philip or Peter who were the disciples of Jesus. Peter and Philip's names are used as the books' titles in the same way as the authors of the Synoptic Bible, the Holy Bible, as already mentioned elsewhere in this work.

Someone else is reporting conversations which took place between Mary Magdalene, Jesus, and other disciples; so like all biblical news, it is up to us to judge the validity of the news. Again it should be clear that these biblical narratives were written during the second and forth centuries and not before that period; because prior to the second century there were no such people or organized group of followers known as Christians. There were followers or churches of the individual disciples of Jesus. These followers were not united but rivals who claimed to be teaching the true Gospel of the Christ's disciples. Christianity as a unified title embracing all the followers of Christ applied after AD 325 when Emperor Constantine and Council of Nicaea brought together and institutionalized the religion under a uniform name currently called Christianity.

The book of 1 Corinthians 7: 12 and other places in the New Testament have used the word 'Christians' a few times in its narratives, suggesting that, the book and all the other Bible books were compiled after Constantine's era. Hence the scriptural texts we read today are carefully cooked, scrutinized, and edited to ensure long-lasting ambiguities.

The Gospels of Mary Magdalene and that of Philip are testifying that there was a sexual relationship between Jesus and his lover Mary of Magdala; and who are we to argue about its validity? There is absolutely nothing in the canon, Judea or Hebrew, which says Jesus Christ ought not to marry a woman.

Modern-day Christians find it extremely difficult even to imagine that their Savior Jesus Christ ever used his penis for his lover Mary Magdalene or any other woman; even a mere lip to lip kisses is seen as unbefitting. They *want* to believe that Jesus only used his penis for urinating, while the so-called modern Christian fellowship cannot let a day pass without sexual intercourse with their lovers. If all pastors, deacons, and religious leaders believe they emulate the biblical image of Christ, why do they marry and have sex with their lovers, and even sometimes have extra sex outside their marriages? No doubt they know very well that Constantine and his council of bishops have deliberately inserted '*Psalm 51*' and several biblical quotations designed as *Forgiveness Prayers* to protect them if they intentionally sin. This endemic hypocrisy and blatant lies in Christianity is gloriously cherished even today as we read this discourse.

The above quoted narratives by Philip have obviously been censored by the ancient clergy to sanctify or enhance Christ's celibacy and immortal image. Not so long ago, a movie was released in the United States called *The Last Temptation of Christ* in which Jesus Christ was having sex with his lover Mary Magdalene and the whole Christian world went crazy. Time has changed, captive minds are getting liberated, and full explanation of biblical mythology and ambiguities is needed.

One of Philip's quotations above, need to be examined closely: *The Gnostic Bible; (The Gospel of Philip), page 313:*

> "There is the earthly son, and there is the son of the earthly son.
> The Lord (Jesus) is the earthly son, and the son of the earthly son is he who is created through the earthly son."

You may be the exception if you are not baffled by Philip's statement. Because if we are to rewrite the sentence in modern wording, the second sentence would read like the following: Jesus is the son of God, but the son of Jesus is the one who Jesus himself and a woman have conceived and brought forth with. *'There is'* a 'positive existence' of Lord Jesus and a 'positive existence' of the son of the Lord Jesus. This is very characteristic of biblical texts, which are intended to be concealed, yet somehow the authors feel obliged to mention with caution that it may not be construed as religious blasphemy; such statements are always ambiguous. If there is Jesus' own son, who is he, who is the mother, and what is the son's name? In the interest of Jesus' mystique holiness and his blessed second name (the Christ), his sexuality will forever remain a secret to the general public. If you, however, read between the lines, you sense out that the man, Jesus, impregnated a woman, possibly his lover Mary Magdalene and subsequently fathered children, at least a son and a daughter with her. In Mark 5; 34: *"Jesus said to her, my daughter your faith has made you well. Go in peace and be healed of your troubles"*. From Gnostic source, Jesus fathered five children; the last child was believed to be called Sara, daughter of Mary Magdalene, who was alleged to be pregnant at the time Jesus was on the cross.

The next sentence that follows this secret father and son sentence is the following which in my opinion need not have been mentioned at all.

'*No one can know when the husband and wife have sex except both two.*' No sex partners, particularly husband and wife, will consciously go about bragging to have had sex, for instance, the previous night. I guess Philip is trying to tell readers that Lord Jesus cannot be explicit with his sexual relationship with his wife Mary Magdalene. If this is so, we will all be in agreement with Jesus as a human being, because most married couples would certainly do exactly the same thing. But if Jesus is the Christ, the divine entity, who according to John pre-existed before creation took pace, I mean *Christ* as the God himself, then I would have a big problem with his sexuality. Would he sleep with his wife Mary Magdalene spiritually as angel Gabriel did, with his own mother Mary; or he would sleep as *Jesus*, the flesh and blood circumcised Jewish masculine man, with proper erection and penetration.

I hope the reader gets my point. You see, the essence of this entire discourse is a search for the existence of The Omnipotent God for the whole universe. If biblical scriptures are telling the whole world that the biblical God is the Cosmic God the creator of mankind, the question of wife, sex, and children do not apply.

The other side of the coin, however, can justify a good case for Jesus Christ to have a constructive sexual partner as according to scriptures without question. For instance, if we can accept the Holy Bible simply as a Book of Fictions, and not different from any other fiction book, I suppose all the characters mentioned can be made to play any role, be it a God or the God, the Messiah or the Christ, Tao or Hindu, Buddha or God of Israel, or whatever character the author will assign will be irrelevant. A fictitious story will always remain one no matter how old or how much anyone manipulates its contents. It can never be real.

The Jesus Christ story in the Bible has already been exhausted in the previous chapters. But briefly, let us look at following: the guy's mother became pregnant mysteriously, by Angel Gabriel (*a fiction*). The mother delivered baby Jesus while still a virgin, (*a fiction*); nothing is written about Jesus' whereabouts between the week he was born and circumcised, and up to the age of eleven, the missing years (*a fiction*). Bar Mitzvah celebration is a traditional requirement of every male Jew who reaches the age of twelve; the Bible mentions it. But nothing about Jesus is written anywhere in the Bible between thirteen and twenty-nine, the missing years (*a fiction*). Age thirty is mentioned as when Jesus began his Ministry (*a fiction*). The ages thirty-one and thirty-two are missing (*a fiction*). His death, mysteriously preplanned by God himself (*a fiction*); three days after death, as according to the plan, Jesus comes alive again in full blood, flesh and bone, like a perfect normal human being (*a fiction*). Jesus finally vanished to heaven to stay with his Father (*a fiction*), yet he reveals himself spiritually to his disciples occasionally (*a fiction*). There is a predicted judgment day when Jesus will return the second time to earth and go back to heaven with righteous people (*a fiction*).

Everything about this creature is based on the Old Testament prophesies which are primarily *fictitious* and historically very loosely based. If Jesus Christ is a fictional character, which I am very convinced he is, then who cares whether he was married or not, or how many women he slept with daily. *He cannot be real and remain unreal,* at the same time. It is illogical. Jesus Christ's dual personality (the human Jesus and the spiritual Christ) make the guy resemble a phony theatrical character. And I am certainly not alone to draw this conclusion. It is this very coexistence of these two personalities which brought about the various divisions in Christianity. Several denominations right from the first Bishops' Council Meeting, 325 years after Jesus' death, as already discussed under the Council of Nicaea, tried unsuccessfully to merge them.

It was this very issue that brought into existence the Eastern Orthodox Church, the Roman Catholic Church, the Lutheran Church, the Episcopal Church, and the Church of Jesus Christ Later-day Saint which still believes the humanity of Jesus is separate from Christ the son of the Heavenly Father.

CHAPTER 11

Questionable Biblical Notes

The biblical Christ according to the book of Isaiah 7: 14, *'a young woman who is pregnant will have a son and will name him Immanuel.'* Note here that Isaiah did not say *a virgin*. Again, *'will have a son'* as written above can mean adopting a son, or possibly conceiving a son regardless of the mode of the conception. Evidently, this is one of 1001 biblical ambiguities. And needless to say, it helps to wonder about the real *holiness* of the Holy Book itself.

Isaiah 11: 10, *'A day is coming when the new king from the royal line of David will be a symbol to the nations.'*

'(The) new king.' Which new king? Does this refer to Jesus the Christ who was never a king or was he ever a crowned king during his secluded years between the ages of thirteen and twenty-nine? Which hereditary monarchy did Jesus inherit? Was it the Pontius Pilate's mockery of Jesus during the Passover celebration when he asked whether the Jews were referring to 'Jesus the king of the Jews' (Mark 15, 9) or 'Jesus Barabbas' (Matt. 27: 16)? Holy Bible is mute on this issue.

Now *'the royal line of David'*; Was Jesus a descendant of David, or was Joseph not the surrogate father of Jesus? How about *'a symbol to the nations'?* Is the author referring to a symbolic sign, an icon, or a representative to *the* nations of Israelites, Moabites, Canaanites or what?

These prophesy was made over 900 years before Christ was born. However, over seventy years after Christ's death, or if we add Jesus's thirty-three years' life on earth, it adds up to become about 1000 years after this prophesy when the book of Matthew (the first book of the New Testament) came out with the following: Matthew 1: 18-24; *'This was how the birth of Jesus Christ took place.'* (Be reminded that according to biblical history, this Matthew was not the tax collector Matthew who followed Jesus Christ during his lifetime. 'Matthew' is just a name for the book. This Matthew is therefore not an eyewitness to what he is saying. Historically also, someone with the same name claims to have read Marks' account and also heard from other people before he, Matthew, wrote his testimony of Christ.)

> His Mother Mary was engaged to Joseph, but before they were married she found out that she was going to have a baby by the Holy Spirit. Joseph was a man who did what was right, but he did not want to disgrace Mary publicly, so he made plans to break the engagement privately. While he was thinking about this, an angel of the Lord appeared to him in a dream and said Joseph, descendant of David, do not be afraid to take Mary to be your wife for it is by the Holy Spirit that she has conceived. She will have a son and you will name him Jesus-because he will save his people from their sins. Now, all this happened in order to make what the Lord had said through the prophet come true. 'A virgin will become pregnant and have a son, and he will be called Immanuel' which means God is with us. So when Joseph woke up, he married Mary, as the angel of the Lord had told him to do. But he had no sexual relations with her before she gave birth to her son. And Joseph named him Jesus.

The primary necessary question is why it is only the Gospels of Matthew and Luke who account for the birth of Jesus amongst the entire sixty-six books of the Holy Bible?

Indeed there are multiple question marks about the above quotations: for instance, Isaiah's prophesy did not use the word 'virgin'. Please read the quotation above. The author of the book of Matthew is deliberately misquoting Isaiah. Isaiah refers to a 'young woman', or a female youth who is not necessarily a virgin or one who has never had sex or a child prior to Joseph's daydream. Matthew's author has indeed expressly stated that 'in order to *make* what the Lord had said through the prophet *come true*,' Mary *had to be* a virgin *to fulfill* Isaiah's prophesy. Yes. Again note also that it is Joseph, *not* Mary, who is of David's ancestry. If Joseph did not have sex with Mary *before Jesus was impregnated*, it means *Jesus's blood link with David is zero and therefore Jesus could not be the Isaiah's prophesized 'symbol'*.

Jesus did not have Joseph's blood in him by birth, unless of course there was sexual intercourse between Mary and Joseph prior to Mary's pregnancy. Again the word 'symbol' is used to describe the destiny of Mary's son in Isaiah 11: 10. The word 'symbol' has at least eight meanings and none is related to a Messiah. The closest synonym is probably 'an icon'. Firstly, Isaiah did not say 'a messiah' or it may be by implication, but even then, the Jesus biography in the Holy Bible does not qualify him to be the prophesized messiah.

It is very tempting to put words into Isaiah's mouth, and I think every critical reader ought to be careful. Biblical stories are noticeably and fundamentally mythical, like all myths, dramatisation of scenes such as the birth and death and resurrection of Jesus Christ is very crucial and indeed must be told with extreme caution. Really we don't have to be particularly smart to see the logic in the story.

Before Jesus Christ's era, there was a mythical story—the old Egyptian myth of Dionysus' virgin birth by Isis. The virgin story appears to have been *copied* for Jesus Christ's birth by New Testament authors. Isis, the virgin mother, spiritually became pregnant, even though her husband Orsis did not have sexual intercourse with her. As in the Bible, Dionysus became Jesus, Isis became Mary, and Orsis, Joseph; the equation is perfect. It enhances the mystery surrounding the lead actor who at this point is Jesus the Christ who is prominently portrayed in the New Testament books as God incarnated. Some Christians conclude that God used Mary as a tool to create Jesus. Being the all powerful entity, God could have made Jesus come to being by any way he chose, but just to fulfill biblical prophesy, his birth must be made through a young woman. Jesus therefore is Isaiah's fictional character, and Isaiah's followers among the Jews wish to see that the character becomes a reality. God of Israel therefore had nothing to do with Jesus's birth.

The author of the book of Mark was concerned about Christ's supposed deeds—his acclaimed magical works, his portrayed righteousness, his death, and his legendary resurrection. Biblically, Christ lived, died, and was resurrected in both human form and in spiritual form. God was his father. How he was conceived and born was not important to Mark. Jesus ascended to heaven to live with his father-God. He, however, promised to return soon to have everlasting life with the righteous people of the universe.

Jesus, the prophesized Christ, would soon come when both the dead and the living are called before God for judgment. About thirty to forty years after the death of the biblical Christ, Paul—the leading architect of Christ's godly image—started drafting any conceivable divine character for the man. He made the human Jesus become the Christ, the divine or the Son of God, the Holy Ghost, and God the Father himself—The Holy Trinity. Islam not only detests Pauline's characterisation of Jesus as the Christ, rather Muslims recognize biblical Christ *only* as a prophet of God when Jesus was alive.

According to the Holy Quran, Jesus died like any mortal human being and was never resurrected. Gnostic scriptures and recently revealed history supports this Islamic tradition. Similarly, Hebraic scriptures or orthodox Jews do not recognize Jesus Christ as the prophesized Messiah. The Jewish son of Yahweh, their God, is not born yet. According to this faith, the biblical Jesus Christ is counterfeit. This is the major reason why Jesus' own ethnic people (Jews) arrested him and had him killed. Rabbinic Judaism is still waiting for the Christ to be born according to Isaiah's prophesy quoted above. It should also be noted that Bible scholars are blamed by Orthodox Jewish scholars for mistranslations or misunderstandings of the Hebrew texts.

Some Christian denominations such as 'The Church of Jesus Christ of Later-day Saints' believe that God and Jesus are literally separate persons. The Holy Trinity is not acceptable to their faith. The only religious faith which still believes the authenticity of the biblical Christ as the Messiah and the son of the Jewish God (the same Jewish Yahweh) is 'Christianity'. Unfortunately, however, *there is no conclusive historical evidence to support the little biography of the biblical Christ in the Holy Bible.* There is, however, overwhelming historical evidence that supports the assertion that the entire biblical saga is a fiction and nothing but hoax.

The book of Exodus for instance attempts to narrate the movement of Jews from Egypt to their God's promised land, which was already occupied by several tribes according to the author. Several eminent archeologists and historians have devoted their lifetime to search for facts about the route, settlement areas, people's names, tribes, towns, cities, rivers, deserts, distances, ages, and times plus several pertinent references; but no credible conclusion absolutely supports the biblical texts. (Refer to bibliography; M.D. Coogan, W.T. Pitard, C.A. Redmount, and L.E. Stager).

Theologically, *some* biblical moral texts may be important for social purposes, but historically they are inaccurate, irrelevant, and ought to be ignored. Where is the biography of the biblical Jesus Christ in the entire sixty-six books of the Holy Bible? His life story is curiously missing from the book. Readers are merely informed of four episodes of the man; his birthday, his twelfth birthday, his thirtieth year, and his thirty-third year when he was allegedly crucified. Yet readers are informed this by the author of John at the conclusion: *'Now there are many other things that Jesus did. If they were all be written down one by one, I suppose that the Whole World could not hold the books that would be written'* (John 21: 25).

A thirty-three-year-old person's biography when completely written down, the entire universe could not contain the number of books? Come on, give me a break! How on earth can this personality, Jesus Christ, not be a fictional character is beyond belief. Some of us may see no need to question this foolishness in the last sentence above, but certainly not all of us. How many books can possibly be written—a million, a hundred million, or a billion books? How's it possible? Or is the book of John referring to biographies of all the three Jesus in the Holy Bible together, *i.e.* Bar-Jesus (Act 13; 6), Jesus Barabas (Matthew 27; 16), or the Jesus in Mark 10: 18, who refused to be called 'good' because '*No one is good except God alone*'. This guy (Jesus Christ) was either a crook, or fictitious, hence his entire biography must be kept as a secret. When his full story is told, there is no way he will retain the holiness title assigned to him.

The last gospel of John 21: 25 certainly gives an obvious impression that Jesus Christ was probably a crook. Hence his entire biography need not appear in the Holy Book.

Or that his 'holiness' will be diminished if his human story is fully exposed. 'To err is human'—so says the Bible. Hence if the biblical Jesus Christ was a real human being who lived for thirty-three years on planet Earth like you and I, he must have committed some sins, or probably actually made too many human errors that deserve to be hidden.

Ancient theologians deliberately erased Christ's biography, particularly his faults, to avoid embarrassments. I am inclined to visualize ancient peoples' understanding or definition of the universe as used in the book of John quoted above, the same sense as used in the book of Genesis 1: 6, the universe as a 'Dome'.

'He named the dome 'sky' (Gen. 1:8).

I am sure in primitive days a hamlet resident across river Jordan could look up into the sky and measure the end of the universe with the horizon; just a couple of miles forward, behind, left and right, up and down and assume a perfect knowledge of the globe, a miniature world. Even so, how many books can we suppose a hamlet of ten houses cannot contain. It is doubtful if the author of the book of John really thought matured readers, like you and I, would accept the book's conclusion without questioning the logic. An immature five-year-old child will certainly not believe this biblical nonsense.

A major questionable biblical text is found in the book of Hebrews 1: 1-3:

> In the past, God spoke to our ancestors through the prophets, but in these days he has spoken to us through his Son. He is the one trough whom God created the universe, the one whom God has chosen to possess all things at the end. He reflects the brightness of God's glory and is the exact likeness of God's own being, sustaining the universe with his powerful word.

The book of Hebrews was written around eighty to ninety years after Christ had died. And we are told that Christ lived only thirty-three years. God had created this universe several millions of years before Christ and his mother were born. Again this author could not claim to be an eyewitness to what he or she is writing about. How did he know that God created the universe through Christ? In the past, God spoke to our ancestors through the prophets. These days God speaks to us through his Son. Whose ancestors and which days? Jewish ancestors and the year 2011 days? Should we not ponder about these malicious statements?

Again if we look into Gnostic scriptures, a whole different picture of this guy (Jesus) pops up. Such books as *The Banned Book of Mary (Jesus' Mother)*, *The Gospel of Mary of Magdala*, *Beyond Belief (the Gospel of Thomas)*' *The Gospel of Philip*, *The Gospel of Barnabas*, *The Templar Revelation (Guardians of the True Identity of Christ)*, *Holy Blood Holy Grail*, *The Gnostic Bible*, and many other serious research works reveal most of the Holy Bible puzzles, especially the Synoptic as opposed to Gnostic scriptures in the New Testament.

There exist the other biblical stories besides those in the Holy Bible and it is very important to be curious and also to search for the untold story, than to bear false witness to allegations which common sense clearly rejects.

The one book which interests me most, and which I think must be isolated from all the sixty-six books need be zoomed closely for better perspective of the cosmic God we are searching for: that book is 'Ecclesiastes;' the twenty-first book of the Old Testament.

Chapter 12

The Book of Ecclesiastes

Ecclesiastes is a name or a word which stands for a teacher or a preacher. Its equivalent in Hebrew translation is '*Qoheleth.*' It is one of the books of the Hebrew Bible. The main speaker in the book identifies himself as the 'teacher' or the 'preacher' who is the son of David and also the King of Jerusalem. Some Hebrew scholars believe the author of Ecclesiastes was King Solomon.

The book contains twelve very short chapters with personal or autobiographic matters reflecting the meaning of life and the best way to live it.

The author writes that human life on planet earth is inherently vain, futile, empty, meaningless, temporarily, transitory, fleeting, and mere breath. He intermittently mentions God's name in the same way as the God portrayed in the book of Genesis. He however assigns every earthly event to another biblical God character that resembles the God of Israel as exhibited in the books of Exodus, Leviticus, Numbers, Deuteronomy and Joshua.

Nevertheless, his perception of life in general is perfectly understandable, considering the quality of human knowledge four hundred years before Jesus Christ was born. I am absolutely certain that if he was alive today, his analysis of human life would be identical to our present discourse. Indeed if all of us were alive then (fourth century BC) we might probably have Ecclesiastes' view of life.

This is AD twenty-first century. Things are different. Humans possess quantity and quality knowledge of our cosmic world. For the sake of convenience, let us examine the following excerpt from the book. This is from Ecclesiastes 1; 2: 10:

> It is useless, useless, said the Philosopher. Life is useless, all useless. You spend your life working, laboring, and what do you have to show for it? Generations come and generations go, but the world stays just the same. The sun still rises, and it still goes down, going wearily back to where it must start all over again. The wind blows south, the wind blows north—round and round and back again. Every river flows into the sea, but the sea is not yet full. The water returns to where the rivers began, and starts all over again. Everything leads to weariness—weariness too great for words. Our eyes can never see enough to be satisfied; our ears can never hear enough. What has happened before will happen again. What has been done before will be done again. There is nothing new in the whole world. Look, they say, here is something new! But no, it has all happened before, long before we were born.

Up to this point everything the philosopher talks about is perfectly true about human life. But it is never true that 'there is nothing new in the whole world. Look they say here is something new' *Yes, I say it is true; there can be something new in this new world we have lived in since the past 100 years.*

Since the early AD 1400, when the Catholic religious status quo began to crumble from within its institutional base, many things which were previously unquestionable because they were deemed God's prerogatives have become part of our everyday investigation topics. Philosophical ideas may not be completely new, because it may be a splice of an existing philosophy; but even then as long as it is an accepted spliced idea; it necessarily becomes a legitimate new notion. Again, invention of new material things in the modern religiously liberated world is part of everyday phenomenon. Hence Ecclesiastes' statement that 'it has all happened before, long before we were born,' is part of the usual biblical cheap narratives.

> No one remembers what has happened in the past, and no one in days to come will remember what happens between now and then.
>
> I, the philosopher have been king over Israel in Jerusalem. I determined that I would examine and study all the things that are done in this world. God has laid a miserable fate upon us.

Is there such a thing as fate? If there is destiny, isn't it our own making, a product of the human brain? Why this last statement 'God has laid a miserable fate upon us'? After a second thought, I withdraw my question. Because, at the time this thinker was writing his observations, everything that was not explicable by human intelligence then was assigned to God. Primitive minds in the past always resort to divine or superstitious source for answers to every earthly mystery.

> I have seen everything done in this world, and I tell you, it is all useless. It is like chasing the wind. You can't straighten out what is crooked; you can't count things that aren't there. I told myself.
>
> I have become a great man, far wiser than anyone who ruled Jerusalem before me. I know what wisdom and knowledge really are.
>
> I was determined to learn the difference between knowledge and foolishness, wisdom and madness. But

I found out that I might as well be chasing the wind. The wiser you are the more worries you have; the more you know the more it hurts.

This is from Ecclesiastes 2: 1-10:

> I decided to enjoy myself and find out what happiness is. But I found that this is useless, too. I discovered that laughter is foolish; the pleasure does you no good. Driven on by my desire for wisdom, I decided to cheer myself up with wine and have a good time. I thought that this might be the best way people can spend their short lives on earth. I accomplished great things. I built myself houses and planted vineyards. I planted gardens and orchards, with all kinds of fruit-trees in them. I dug ponds to irrigate them. I bought many slaves, and there were slaves born in my household. I owned more livestock than anyone else who had ever lived in Jerusalem. I also piled up silver and gold from the royal treasuries of the lands I ruled. Men and women sang to entertain me, and I had all the women a man could want. Yes, I was great, greater than anyone else who had ever lived in Jerusalem and my wisdom never failed me. Anything I wanted, I got. I did not deny myself any pleasure. I was proud of everything I had worked for . . .

'In days to come we will all be forgotten. We must all die—wise and fools alike' (Eccl 2: 16).

> Nothing that I have worked for and earned meant a thing to me, because I knew that I would have to leave it to my successor, and he might be wise, or he might be foolish—who knows? Yet he will own everything

> I have worked for, everything my wisdom has earned
> for me in this world. It is all useless! (Eccl 2: 18-19)

'You work for something with all your wisdom knowledge, and skill, and then you have to leave it all to someone who hasn't had to work for it. It is useless, and it isn't right' (Eccl 2: 21).

In our struggle to stay alive, some of us with religiously liberated mindset are able to acquire more than we practically need for survival. We acquire and maintain every need in excess. Consequently, our lifestyles become affluent simply because we are more fitting and capable than others who are competing for the same substance. Life is useless only if we do not understand the meaning of birth and death. I am pretty sure the philosopher understands it. He just wishes he could have everlasting life—and that is not life. The essence of life is to procreate to guarantee continuity of human existence. We must die so that our children and children's children will have space and chance to live. But before death we must try to prepare a better place for our children to live with minimum struggle.

> As long as you live, everything you do bring nothing but worry and heartache. Even at night your mind can't rest. It is useless. The best thing a man can do is to eat and drink and enjoy what he has earned.
> And yet, I realized even this comes from God. God gives wisdom, knowledge, and happiness to those who please him. (Eccl 2: 23-24)

There is a contradiction here: compare the above godly statement in chapter 2, verses 19 and 21—also stated above and repeated below here:

'You work for something with all your wisdom, knowledge, and skill, and then you have to leave it all to someone who hasn't had to work for it.' The philosopher has *earned* wisdom, knowledge and skill through hard work, not by chance or by divine means; according verse 19, yet he is linking his accomplishment with God instead of praising himself for his earnest hard work.

Wisdom, knowledge, and skill are the dependable total sum of quality accumulated information stored in the human brain. What is the supporting evidence that these three virtues were given by God. There are no miracles in hard work. Every hard work has its unique proportionate reward. The harder you work the better your reward. The less hard work you do the less reward you obtain. That is the life we choose to live. No God is responsible for our failures and successes. The philosopher's success has nothing to do with God. It is the result of his wisdom, his knowledge and skills. I think he should rather advise his readers to stay focused to their dreams in pursuit of prosperity, good health, and happiness. Dependence on God can thwart human efforts.

Ecclesiastes 3: 1-2: 'Everything that happens in this world happens at the time God chooses. He sets the time for birth and the time for death; there is even *'a time to kill' in verse 3.*

On questions of birth and death of living organisms in general, and indeed in most scientific observations, ancient people typically always explain them in philosophical terms. What we understand today as natural sciences in general, and in particular, medical science, organic anatomy, and physics was part of ancient philosophical studies. For this reason, the philosopher cannot be criticized. There is no evidence or can ever be any evidence that birth and death decisions are preplanned by any divine entity called God. These events are man-made. A decision to give birth to a human being is dependent on choices made by human beings; and more so by the female career of the pregnancy. She is the sole individual whose decision can affect growth of the fetus to its full maturity term. This has nothing to do with God.

Again, with regards to death every matured individual can decide when to die. Death is a human personal choice. This has always been the case from time immemorial. Anybody can choose to die any year, any month, any week, any day, any hour, any minute and any second. All of us can choose to die whenever we please; either by sudden suicide death or by gradual bad habit or self-induced careless death. We merely simply instinctively choose to live as long as possible. Death is not, and cannot be an act of God. Just think about it yourself.

Chapter 3, 12-13; all we can do is to be happy and do the best we can while we are still alive. All of us eat and drink and enjoy what we have worked for. It is Gods gift. God will not and cannot, put food or drink into your mouth as a gift or for any reason. You may choose to worship him till their kingdom come, food and drink cannot fall from heaven. It is religion which has planted these illusive ideas into our minds.

Chapter 7: 20: There is no one on earth who does what is right all the time and never makes a mistake.

Yes! This is truism. And I am pretty sure this includes Jesus the Christ: that is if we admit that he ever lived on planet earth as a human being like you and I.

Verse 29; God made us plain and simple, but we have made ourselves very complicated. Chapter 10, 14: no one knows what is going to happen next, and no one can tell us what will happen after we die. The last chapter and the last verse: *Chapter 12, 13-14: 'After all this, there is only one thing to say: Fear God, and obey his commands, because this is what man was created for. God is going to judge everything we do, whether good or bad, even things done in secret.'*

Judging by the last two verses of the book of Ecclesiastes, as quoted above, the philosopher's statements about God resonates to the God of Israel as revealed in the book of Exodus through to the book of Joshua. And if this is the God who is going to judge everything we do on planet earth on the dooms day, I guess all of us should make it our duty to revisit these five books without delay: because by the narratives in those five books, that God of Israel is not fit to judge anyone or anything on this planet. I challenge everyone to read the books for personal conclusions.

So far, the Apostle Paul whose deeds as a living human being, and whose role in Christianity is next to Christ, has not been discussed yet. It seems appropriate now to examine his position with regards his apostolic image of the cosmic God we are deliberating on.

Chapter 13

Saul: the Apostle Paul

Saul was born five years after the death of Jesus Christ. He was born in Tarsus, a south central town of in Turkey. His mother and father were both Jews. His native country, Turkey, was part of the Roman Empire; hence he was a Roman citizen. Legally however Roman, citizenship at the time was not an automatic right; it was an honor which was sold to deserving citizens. Paul's parents had already purchased the family's citizenship right. He spent his adult years with his parents in Jerusalem. He was beheaded and died in Rome in AD 67.

By the age of thirty-one to thirty-six, Paul was a pagan and known as Saul. *Acts 13: 9 says, 'Saul also known as Paul'*. This is what the book of Acts writes about him.

According to Acts 22: 1-5

Brothers and fathers, listen to me as I make my defense before you.' When they heard him speaking to them in Hebrew, they became even quieter and Paul went on. 'I am a Jew, born in Tarsus in Cilicia, but brought up here in Jerusalem as a student of Gamaliel. I received strict instruction in the Law of our ancestors and was just as dedicated to God as all of you who are here today. I persecuted to the death the people who followed this way. I arrested men and women and threw them into prison. The High Priest and the whole Council can prove that I am telling the truth. I received from them letters written to fellow Jews in Damascus, so I went there to arrest these people and bring them back in chains to Jerusalem to be punished.

Verse 6-16

As I was travelling and coming near Damascus, about midday a bright light from the sky flashed suddenly around me. I felled to the ground and heard a voice saying to me, Saul, Saul! Why do you persecute me? Who are you, Lord? I asked. I am Jesus of Nazareth, whom you persecute. He said to me. The men with me saw the light, but did not hear the voice of the one who was speaking to me. I asked what shall I do, Lord? And the Lord said to me, get up and go into Damascus, and there you will be told everything that the God has determined for you to do. I was blind because of the bright light, and so my companions took me by the hand and led me into Damascus. In that city was a man named Ananias, a religious man who obeyed our Law and was highly respected by all the Jews living there. He came to me, stood by me and said, Brother Saul, see again! At that very moment I saw again and looked at him. He said, the God of our ancestors has chosen you to know his will to see his righteous Servant, and to hear him speaking with his own voice. For you will be a witness for him to tell everyone what you have seen and heard. And now why wait any longer? Get up and be baptized and have your sins washed away by praying for him.

Verse 17-29

> I went back to Jerusalem, and while I was praying in the Temple, I had a vision in which I saw the Lord as he said to me, hurry and leave Jerusalem quickly, because the people here will not accept your witness about me. Lord, I answered, they know very well that I went to the synagogues and arrested and beat those who believe in you. And when your witness Stephen was put to death, I myself was there, approving of his murder and taking care of the cloaks of his murderers. Go, the Lord said to me, for I will send you far away to the Gentiles.
>
> The people listened to Paul until he said this; but then they started shouting at the top of their voices, away with him! Kill him! He is not fit to live! They were screaming, waving their clothes, and throwing dust in the air.
>
> The Roman commander ordered his men to take Paul into the fort, and he told them to whip him in order to find out why the Jews were screaming like this against him. But when they had tied him up to be whipped, Paul said to the officer standing there, 'Is it lawful for you to whip a Roman citizen who hasn't even been tried for any crime?' when the officer heard this he went to Paul and asked him, 'tell me, are you a Roman citizen?' Yes, Paul answered. The commander said, 'I became one by paying a large amount of money.' 'But I am one by birth' Paul answered. At once the men who were going to question Paul drew back from him; and the commander was frightened when he realized that Paul was a Roman citizen and that he had put him in chains.

The story of Paul is widely spread throughout fourteen of the twenty-seven books in the New Testament. The central theme of Paul's Christology is faith in the biblical Christ as the true son of the God of Israel. And it is only through his grace that our sins can be forgiven. During his lifetime his dream was not realized by linking the traditional Judaic Hebrews with the ordinary Jews or uncircumcised Gentiles as well as Jesus Christ followers into a common theological worship of a single God through Christ. It was a decade or so later after Paul's death that other gospel stories emerged, beginning with Mark, Matthew, Luke, and John according to 2nd biblical writers already discussed.

Indeed there were other significant Christians such as Peter, James, Simon, Philip, Mary Magdalene, Mark, Thomas, Barnabas, John Mark and several others whose Contributions have been marginalized by posterity. But however Paul's theology we might accept or reject, one vital point is that faith is nothing but a product of the human brain. It is dependent on intuition. Every human culture creates its own unique faith base as an engine of cohesion and survival. The essence of faith is defused when indigenous faith is replaced with another culture's faith.

Christianity is simply a Jewish faith, and should be retained as such, if pursuit of external faith is anyway or anyhow necessary. Faith of the 'self' is the ultimate human need; certainly not faith beyond the individual self.

CHAPTER 14

Conclusion

I can only hope that the singular message I am trying to communicate with my readers is written clearly and well expressed for better understanding of the message. Briefly, however, my message is this: 'a single cosmic God is an absolute impossibility and non-existent'. That, a single entity called God could not be the designer and controller of planet Earth and its content. The God, as narrated in the Bible, cannot be the creator of anything. The proposition written in the Holy Bible is too weak and lacks verifiable evidence. You can, however, choose him to be your God. And I am not concerned about your individual choice. My concern is that I have also read the Bible very carefully and the three main god images described in the book; 'the Creator God', 'the God of Israel', and 'the New Testament multiple Gods', do not have convincing evidence to qualify to be what they claim to be. All of them fit perfectly well into classic definition of fictional characters. You cannot merge the three Gods either to become a single God, as the Holy Bible authors attempt to do, in any logical sense.

Anybody can of course create God's imagination in the mind as indeed many millions for millions of years have done before, but remember also that these perceptions will forever remain wishful thinking and a mere fantasy in people's mind.

Our world, the planet Earth as a single entity, is too complex and probably the most complex among other known planets. It is such that it cannot have come into existence by a single genius creature called God by any miraculous means, especially as narrated in the Holy Bible. Some thinkers use the planet's complexity as a good case for God's existence. They claim that the world is too complicated to have come into being without a super intelligent entity behind it. This proposition is OK as long as the proposer is capable of supporting this wishful idea with indisputable evidence. But this argument defeats itself. If for instance, we limit our scope of inquiry to the existing *visible* living organisms alone, the inherent multiple interactions and coexistence of billions of creatures in our solar system make the idea of a single God unimaginable.

If we turn our minds to the micro world of *invisible* living organisms, the picture becomes more perplexing. The invisible organic bacteria world, which is not visible to the naked eye, and which coexist with larger organism like you and I, the living together of micro and macro organisms in the same cosmic universe, is surely amazing. It will definitely be a defeatist attitude for you and me to simply conclude that, just because of the world's complexities, someone mysterious must have engineered our planet. We will be defeating ourselves to simply say that God created them without verifiable supporting evidence. This perception indirectly prevents anyone to think about it. It presumes that mankind actually has the answer to the question of our source. The fact is, you do not know for sure and I do not know for sure, so we must be careful to say that we know it just because 'a' book (the Holy Bible) says so. How do you know? Is it just because the Bible says so? Let's not forget that the Bible is a man-made object, a book, like million other man-made things. As already explained elsewhere, the Holy Bible was literally planned, word by word, by ordinary human beings like you and me. The

Bible was not written by God or by angels. The Bible has its source history. We know how it came into being. It is a codified text of primarily Jewish perverted religion.

All biblical texts are narrated by anonymous authors who are merely reporting stories they have heard about, plus events which occurred decades or centuries prior to the times they were doing the reporting. Once again, we defeat our own intelligence by not looking elsewhere for our makers, when we do not accept the idea that our mothers actually designed us in their wombs. You may refuse to accept this natural fact or you may still think someone else made you and your ancestors, because the Holy Bible says. There is no scientific or verifiable evidence which demonstrates the existence of god or angel in our mothers' womb during the pregnancy period. For those who are still in doubt of this observation, the best advice is, 'Go and re-read Genesis 1 to Genesis 2: 26 very meticulously this time, and think about it.' Then after that, read the ensuing six books to the book of Judges, and once again, very carefully, and I bet you will encounter a completely different God. We should not just accept biblical stories as holy truth, just because it is so written in the so-called sacred book.

The New Testament man-made God (Jesus the Christ) is fictitiously narrated as an impossible character. His theological role when compared with Genesis creator God and the God of Israel is dastardly presented.

I am aware that whatever I have said here is not new, and I am not claiming any originality here. I am only espousing and amplifying what I believe to be the real truth after comparing theological ideas in the Holy Bible and of the past and the present world experiences as we know it today.

The second message I am trying to communicate with my readers is *the question of faith*. The Holy Bible contains a lot of useful books and very pertinent ideas regarding faith.

My perception of faith and that of the Holy Bible, in general, are the same. The only difference is the elements of faith where we differ. The Bible teaches faith in God or Christ and its benefits. It is concerned with divine or ideal faith. But I speak of practical verifiable faith. Not faith in anything outside the individual human self. With my idea of faith, there is no need for God and the end result is splendid, predictable, and real. I see that most people are in denial of their inner capabilities, especially religious people. They waste useful time, money, and other resources hunting for extraterrestrial help and refuse to create faith or confidence in themselves. They fail to realize that the brain organ in the head is much powerful than any imaginary god on planet Earth. When matured human brain is exposed to quality information, nothing is impossible. The only requirement or condition attached to this idea is to nourish the brain with quality information and maintain focus. No need for God. In fact, a resort to God will divert the brain out of focus or a distraction from the brain's intent and purpose. The book of Hebrews, Ecclesiastes, Timothy, and a few other chapters in other books of the Bible have narrated a lot about faith and its connection with the brain. For instance, *Hebrews 11: 1-3 says,* 'To have *faith* is to be sure of the things we hope for; to be certain of things we cannot see. It was by their *faith* that people of the ancient times won God's approval. It is by *faith* that we understand that the universe was created by God's word, so that what can be seen was made out of what cannot be seen'.

My understanding of the above and many such biblical narratives is that faith in human experience is the central master key that opens every door. And I hope my readers will not misunderstand this discourse. Some of my utterances may sound repulsive, but those are not my real intentions. I really mean to engage readers' full attention in respect of my search for the existence of cosmic god. You will be the judge if I failed or not.

My friends often ask, 'If there is no God, no spirit, no ghost, no witches, no angels and demons, what is there for humans to believe in?' I always answer with a question. Must humans maintain belief or faith in anything outside the self at all, especially something related to religion. I don't think so! Faith in God is avoidable.

Faith in anything outside the self is demonstrably not dependable, not predictable and in fact very counter-productive. Religious faith may temporally increase your comfort level, but it freezes the knowledge of the self and therefore prevents humans to face head-on challenges of the practical world. Religious faith exposes human beings' vulnerability to religious fanatics, thus enabling leaders identify followers' weaknesses for maximum manipulation and exploitation of out-of-focused minds in exchange for non-existing spiritual guidance. It encourages total self-submission and hinders self-esteem. Indeed before Martin Luther came to the religious public scene, the Pope at Rome in his capacity as the head of the Catholic Church—the Empire's official religion—applied biblical scriptures to control followers' mind with impunity. There are several historical testimonies to this effect.

Our world, the planet Earth, has been under constant study for several centuries by very curious minds, minds of people like you and I who refuse to accept superstitious description or diagnosis of this world. In our recent past, especially since the beginning of fifteenth century, some curious individuals in some cultures took upon themselves to challenge status quo description of source of life in particular and that of the globe as a whole. With extreme curiosity, they were able to erase religious brain cells which served as a dark screen covering their mindset. With liberated minds, they discovered that our world and the life in it are sustained by certain observed natural laws but not divine laws. Consequently, the source or the true secret of our being has nothing to do with divinity or the Jewish God. With liberated minds, they are able to invent tools and equipments to probe our universe for better understanding. Many global phenomena today are scientifically explicable because a few people with quality mindset or good knowledge of what make us tick have dedicated their life to search for elements of our universe.

Those of us who are still entangled with the ancient view of this world will forever live in darkness, depend on others for survival, and continue to live in decadence and life without progress. Their eyes are closed and hands clapped together everyday reciting the Lord's prayers for guidance. Curiosity of the human brain has opened several fields of study of our cosmic system such that all notions of God's existence are taboos in most progressive cultures of the world. Needless to say, it is those liberated cultures with liberated mindset which are making the world a better place to live in, while those with Godly addictions forever remain dependent on liberated mindset cultures. And I challenge you the reader to look around yourself for a moment and think deep, if anything around you would have been possible with a religiously closed mindset.

Do remember always that your ancestors, who lived before the sixteenth century, knew nothing about the biblical God. The Book is relatively new in most cultures, especially in the so-called Third-World countries. There is no reason why you cannot live progressively without it. The inquisitiveness of those few individuals brought about quality knowledge of the fundamental natural principles governing our planet: principles which are verifiable and proven reliable. I am not a scientist, and I do not have to be one to be able to examine my mindset. I am just an ordinary curious guy.

There are natural laws of physics which describe the physical nature of our universe with multiple complexities. These laws are not man-made or god-made. Curious minds have searched beyond our solar system and found out that the million stars in the sky represent other planets with different physical nature. The nature of planet Earth is what it is, like the nature of other planets. They are what they are. They couldn't have come into existence by magic. There is no evidence whatsoever which shows that some genius entity designed them, or could create them. The intricate complexity of the planet's composition makes it unthinkable for anyone to imagine them coming into being by extraterrestrial power source. Primitive minds like those in the Holy Bible attempt to link the genesis of the world with childish hypothesis and without substantiating evidence.

Physics, biology, and chemistry also describe the nature of organic matter in our cosmic system. It is demonstrable scientifically that this universe consists of billions of self-controlled molecules which relates to each other and also work together in concert for the survival of each other. These are tested observations and have nothing to do with divinity or god.

You may still wonder who made the molecules and the atoms etc. endlessly to infinity. But be assured that the ultimate conclusion of source will rest on your personal mindset.

Astrological sciences have similarly not discovered any domain or residence of God of Israel or Heavenly city for righteous people in the hemisphere. Through this science, we understand that planet earth is not flat as it was believed to be. We know for certain that planet earth is egg-shaped and in constant circular motion; while the moon and the sun remain constant. People on planet earth see day light when the center gravity moves to face the sun. Again we experience night when the earth's rotation moves away from the sun towards the moon.

Astronomers have not yet discovered the Heavenly God in action causing the appearance of either the sun to bring light or conjuring darkness to appear on earth. Laws of nature control the earth's movements and its content. This is an indisputable scientific fact.

My final quest has already been stated in several chapters of this work, but briefly, I like to repeat that the Holy Bible is an ancient document which attempts to explain the origins of planet Earth and the source of all organic matter. It was written at a time when human knowledge in everything was limited to every culture's horizon or environment, consisting of barely a hundred miles radius. The total discoveries made by curious minds in recent years, especially during the past 100 years, or indeed since the 1950s, is mindboggling. These discoveries have rendered all theological precepts as void. Mankind needs no God. But if you, the reader, think you need God for your comfort, please by all means keep your own cultural God. If you worship a God imported from another culture, you and your offspring will become subservient to that culture. What mankind needs is continuous full exploitation of the human brain potential or power to the optimum level to enhance better understanding of mutual coexistence of all planetary organic species, because in reality, nothing can be something without quality knowledge of other things.

Bibliography

1. Good News Bible. Today's English Version. The British and Foreign Bible Society, London. 1986
2. *The Gnostic Bible*; W. Barnstone and M. Meyer. Shambhala Publication Inc. Boston. Mass. US. 02115. 2009.
3. *The Oxford History of the Biblical World* by M. D. Coogan. 1998. Oxford University Press. NY.
4. *Encyclopedia of World History* by P.K. O' Brien. 2000. Facts on File. George Philip Ltd. NY.US
5. *Pagan Christs: Studies in Comparative Hierology*; by J. M. Robertson. 2011. General Books. Memphis. Tenn. US.
6. *Does God Exist?* by Hans Kung. Doubleday & Co. NY. 1980
7. *Biology*, 5th Edition, by Campbell, Reece and Mitchell. 1999. Jim Green. Menlo Park. CA. US
8. *The Four Witnesses* by Robin Griffith-Jones. Harper Collins. NY. US. 2000.
9. *The Da Vinci Code* by Dan Brown. Doubleday. NY. US. 2003.
10. *The Gospel of Mary of Magdala* by Karen L. King. Polebridge Press. 2003. Salem, OR. US.
11. *The Banned Book of Mary* by Ronald F. Hock. Ulysses Press. 2004. Berkeley. CA. US.
12. *Beyond Belief: The Secret Gospel of Thomas* by Elaine Pagels. 2003. Random House. NY. US.

13. *The Templar Revelation: Secret Guardians of the True Identity of Christ* by Lynn Picknett and Clive Prince. 1998. Simon & Schuster. NY. US.
14. *Not In Our Genes: Biology, Ideology, and Human Nature* by R. C. Lewontin, Steven Rose and L. J. Kamin. 1984. Pantheon Books. NY.US
15. *Angels and Demons* by Dan Brown. 2000. Pocket Books. Simon & Schuster NY. US.
16. *The Origin of Species* by Charles Darwin. 2004. Barnes & Noble Books. NY. US.
17. *Darwin's Black Box* by Michael J. Behe. 2006. Free Press, Simon & Schuster Inc. NY.US.
18. *The Grand Design* by Stephen Hawking and L. Mlodinow. 2010. Bantam Books. NY.US.
19. *The Celestine Prophecy* by James Redfield. 1993. Warner Books. NY.US
20. *The God Delusion* by Richard Dawkins. 2006. Transworld Pub. Bantam Press. London.
21. *The Blind Watchmaker* by Richard Dawkins. 2006. W.W. Norton & Co. Ltd. London.
22. *God: The Failed Hypothesis: How Science shows that God does not Exist* by Victor J. Stinger. 2007. Prometheus Books. Amherst. NY. US.
23. *Science and Non-belief* by Taner Edis. 2008. Prometheus Books. Amherst. NY.US.
24. *Irreligion: A Mathematician Explains Why the Arguments for God Just Don't Add Up* by John Allen Paulos. 2009. Hill & Wang. NY. US.
25. *Holy Blood, Holy Grail* by Michael Baigent, Richard Leigh and Henry Lincoln. 1983. Dell Pub. NY. US.
26. *Webster's Ninth New Collegiate Dictionary.* A. Merriam-Webster. 1987. US.

INDEX

A

Aaron (Hebrew high priest)   76, 79, 87-8
Abel (Cain's brother)   41, 65
Abraham (Old Testament patriarch)   17, 28, 41, 65, 70-2, 86, 93, 110-12, 135, 147, 153
Acts   71, 133, 140, 199
Adam (first man)   29, 41, 55, 58-60, 63, 65, 71, 73, 80, 116, 139, 153
Africans   57, 115, 153
alabaster   119
Alexandria   37, 44, 48
Allah   27, 29
ambiguities *see* religious ambiguities
ancient   19, 25, 28, 31, 38, 40, 48, 58, 66, 74, 77, 82, 119-22, 151-3, 206-7
angels   9, 11, 29, 71, 109, 112, 115, 121, 125-7, 135-7, 139, 145, 156-7, 184, 205-6
Anne (Mary's mother)   132, 155
Apologist   32, 127

apostles   32, 130, 134, 140, 143, 164, 171
apostolic   171
appearances   25, 110, 133, 150
art   18
Assyria   38
Augustus Caesar   44-5, 138, 157
see also Julius Caesar

B

Baigent, Michael
  Holy Blood, Holy Grail   189
Baptist see John the Baptist
Barnabas (Paul's companion)   28, 189, 202
belief   20, 45, 188-9, 206, 211
Bethany   119, 125, 165-7
Bethlehem   111, 113-14, 138, 157-8
Bible   10-11, 18, 27-32, 38, 64-5, 73-4, 101, 116-17, 128-30, 134, 148-9, 164-5, 172, 187-8, 203-5
biblical   22, 27-8, 32, 53, 56-7, 67,

74, 77, 101, 106, 113, 147-8, 159-60, 175, 183
bibliography 6, 34, 66, 175, 187, 211
birth 28 *see also under* Jesus Christ
bishop 32, 37, 48, 112, 134
brain 16-17, 20-3, 25, 34, 57, 62, 64, 66, 71, 127, 148, 206
Britain 47
Buddha (philosopher) 26, 181

C

Canaan 80, 89-90, 93, 96-7, 101-2, 105, 107, 118
canonical gospels 127
Catholic 10, 46-7, 192
Chalcedon Creed 49
characterisation 37, 159, 166, 170, 186
Christian 9-11, 155
  churches 36, 83
  fanatics 46
  fellowship 179
  literature 168, 170
Christianity 5, 12, 23, 27-9, 32, 36-8, 43, 46-8, 113, 127, 171, 178-9, 182, 187, 198
Christology 6, 32-3, 37-8, 109, 154, 170-1, 173, 202
Christos *see* Jesus Christ
circumcision 80, 97-8, 139, 153
conclusion 203
Constantine I (emperor) 30, 32, 35-7, 45-6, 96, 178-9
Constantinople 37, 46
cosmic 28, 53, 84, 92, 115, 117, 147, 181, 206
Council:
  of Bishops 38, 128
  of Chalcedon 37
  of Constantinople 37, 92
  of Ephesus 37
  of Nicea 36, 38, 92, 127, 178, 182
  of Rabbis 120
Covenant Box 76
creation 27, 41, 53, 55-6, 58, 60, 62-3, 66, 74, 112, 118, 131-2, 151, 180
creator 12, 16, 21, 25, 53, 62, 66-7, 72-4, 85, 91, 101, 107, 124, 131, 203
cross 83
cultural gods 41
cultures 15, 17-18, 24, 31, 39, 50, 57-8, 79, 82, 117, 136, 147, 153, 168, 207

D

Damascus 71, 200
Darwin, Charles 74, 212
David (king of Israel) 29, 65, 110, 112, 138, 185, 191
  ancestry of 111, 113, 185
  death 19, 22, 28, 34-5, 83-4, 94-5, 121, 123, 142-3, 145-6, 149-50, 153, 172, 185-6, 195-6
demons 9, 11, 19, 129, 166, 206, 212 *see also* devil
Deuteronomy 31, 52, 57, 63, 65, 92-6, 99, 107, 137, 191
devil 85, 115, 166
Dionysus (god of wine) 26, 185-6
disciples 31, 109, 117, 119-22, 125-6, 128, 132, 143, 145, 150, 160, 174-5, 177-8, 181
disobedience 60-1, 83, 85-6
dome 67, 188
drama 56, 58, 73

E

Ecclesiastes 6, 51, 109, 190-4, 196-7, 206
Eden *see* Garden of Eden

Egypt   32, 42, 52, 65, 68-9, 71, 87, 172, 174
Elizabeth (John the Baptist's mother))   135-7, 157
emperor   35, 44-7, 124, 141, 158, 170
Europe   44, 46-9, 90
Eve (first woman)   29, 41, 59-63, 65, 71, 80, 100, 116, 147, 153
Exodus   50, 65, 68-73, 75-7, 84, 88, 96, 107, 117, 121, 187, 191, 197

F

faith   11, 17, 19-20, 23-7, 29, 33, 38, 49, 64, 74-5, 126-7, 133-4, 187, 202, 205-7
fantasy   204
father   10, 19, 21, 27, 29, 32, 36, 45, 47, 80-1, 84, 115-17, 127-8, 143-4, 186
fiction   58, 63, 72, 77, 80, 82, 121, 151-2, 187
France   32, 43-4, 47, 90

G

Galilee   28, 45, 110, 113-15, 120, 126, 133, 136, 138, 142-3, 158, 160-1, 165
Garden of Eden   60-1
genesis   15, 41-2, 52-8, 60-7, 70-4, 84-5, 99-101, 106-8, 112-13, 116, 127-8, 131-2, 147-8, 150-3, 205
Gentiles   136, 201-2
Germany   47, 49, 90-1
ghosts   9-11, 19, 146, 206
global   73, 82
globe   22, 67, 102, 147, 153, 189, 207
Gnostic   180
Gnostic, Bible   33-4, 155, 160, 165, 168, 176-7, 189

Gnostic, scriptures   32, 114, 152, 158, 174-5, 186, 189
God   15-19, 25-9, 33-4, 53-68, 70-82, 85-6, 88-9, 96-105, 107-10, 127-9, 134-7, 147-50, 152-3, 186-92, 202-9
   commands of   54, 80, 96, 117, 163, 197
   covenant of   70, 153
   messengers of   20, 26, 85, 94
godliness   49
gospels   32-4, 38, 55, 86, 110-11, 124, 127-8, 130, 133-4, 138, 148-9, 155-6, 166-7, 174-8, 189
governor   45, 116, 124, 138, 158

H

heaven   9, 19, 22, 27, 30, 67, 110, 114-17, 122, 125, 128, 132, 140, 146, 152
Hebrew   12, 36, 38, 56, 60, 72-4, 171, 174, 178, 199
hell   19, 22, 60, 93, 116
hero   103, 136
history   35, 41, 43, 45, 66, 109, 124, 129, 134, 186, 211
Hitler, Adolf   72, 91, 99
   Mein Kampf   90-1
Holy Blood, Holy Grail (Baigent, Leigh, and Lincoln)   189
Holy Spirit   111, 115, 128, 137, 157, 172, 184
homo-ousios   37
human being   152

I

Irenaeus (Bishop of Lyons)   32-4, 38, 48, 106, 112, 132, 134, 173
Isaac (Abraham's son)   18, 29, 41,

65, 70-2, 93, 147, 153
Isis (goddess)   26, 185-6
Islam   27, 29, 186
Israel   20, 47, 57, 67-8, 77-80, 82-92, 96, 98-9, 101-4, 107-8, 119, 127-8, 145-6, 155-6, 202-3
  land policies of   90, 97
Israel (biblical patriarch) *see* Jacob
Israelite   57, 80, 84-5, 98, 104

J

Jacob (Isaac's son)   18, 41, 57, 65, 69-72, 86, 93, 137, 147, 153
James (apostle)   40, 50, 56, 117-18, 143, 155, 157, 202
Jeremiah (Hebrew prophet)   67-8, 123, 128
Jericho   97-8
Jerusalem   110, 115, 122, 124, 133, 139, 144-5, 156, 191, 193-4, 199-201
Jesus Barabbas   123-4, 142, 183
Jesus Christ   26-30, 111-12, 114-15, 121-2, 125-6, 128-9, 132-3, 139-41, 152-5, 158-61, 165, 172-5, 177-8, 182-8, 198-200
  ascension to heaven of   133
  birth of   110, 138
  crucifixion of   110, 120, 123-4, 128, 130, 140, 143-4, 150, 166
  death of   36
  resurrection of   29, 34, 39, 110, 117, 120, 125, 128-30, 132-3, 143, 150, 153, 168, 170, 172
  sex life of   168
Jewish   28, 30, 39, 69, 72, 77, 86
Jews   27, 45, 66, 86, 89, 91, 102-3, 107, 110-11, 116-17, 119, 122, 124, 136, 139
Joachim (Mary's father)   132, 155-6
John (evangelist)   32-3, 55, 171-2
John the Baptist   34, 115, 117, 128-9, 133, 136, 157
Joseph (Mary's husband)   29, 41, 65, 69, 111-12, 114-15, 117-18, 125, 136, 138-9, 144, 153, 155-8, 183-6
Joshua (successor of Moses)   89, 92, 96-100, 103, 107, 137-8, 191, 197
Judges   101-7, 109, 137-8, 205
Julius Caesar   43-5
Jurisprudence   85, 124
justice   11, 18, 44, 96, 124

K

Kama (Hindu god of love)   26
king   45, 50, 69-70, 94-5, 102, 111, 114, 137, 141, 170-1, 183, 193, 211
kingdom   110, 116, 137, 140, 197
Krishna (Hindu deity)   26
Kyrie   38
Kyrios   34, 38, 67, 113, 171

L

land policies *see under* Israel
Law   80, 116, 119, 121, 140-1, 146, 166, 199-200
Lebensraum   72, 90
Leigh, Richard
  Holy Blood, Holy Grail   189
Lincoln, Henry
  Holy Blood, Holy Grail   190
logos   37, 56, 67, 73, 171-2
lord   34, 38
Luke (New Testament evangelist)   32-3, 127, 133

M

Magdala 165, 174-8, 189, 211
Mariology 6, 32, 37, 170-1
marriage 82, 88, 136, 138, 166, 168, 176-7, 179
Martin Luther 31, 48-50, 207
Mary 6, 29, 37, 55, 112-15, 125-7, 131-2, 136-40, 143-4, 152-8, 160-5, 176-7, 184-6, 189, 211
Mary (mother of Jesus) 34, 37, 55, 83, 111-15, 125-7, 131-2, 136-40, 143-4, 152-68, 171-2, 174-80, 184-6, 189, 211
Mary Magdalene 83, 86, 125, 127, 132, 143-4, 162, 166-8, 174-80, 202
matter:
  organic 18, 74, 209
  sacred 148
Matthew (apostle) 6, 32-3, 40, 50, 108-14, 116-20, 122-3, 126, 128-9, 131, 149, 156-7, 166-7, 171-3, 184-5
Mein Kampf (Hitler) 90-1
messenger 29
Messiah *see* Jesus Christ
mind 16-17, 19-20, 22, 36, 64, 66, 72, 81, 86, 127, 146, 151-2, 176, 203-4, 207-9
mindset 25-6, 74, 207-8
Mithras (sun god) 26, 48, 159
Mohammed (prophet) 26, 29
Moses (Hebrew prophet) 17-18, 26, 28, 57, 68-72, 75-7, 79-80, 82-9, 92, 96-7, 107, 116-19, 121-2, 137-8, 145-6
  satanic duties of 107
mother of Jesus *see* Mary
Muslims 186

N

Nazareth 6, 27, 45, 111, 113, 132, 136, 138-9, 145, 159, 200
New Testament 18, 32, 36, 39-40, 49-50, 55-6, 71-3, 93, 100, 107, 109, 118, 155-7, 164-5, 170-2
Nicaea 36-8, 46, 78, 92, 127, 178, 182
Nicene Creed 32

O

Old Testament 28, 33, 36, 39, 50, 55-6, 65, 67, 92, 96, 101-2, 105-6, 108-10, 112, 118
omni 26, 71, 96, 105, 128
organisms 16, 24, 74, 204
Osiris (Egyptian god of the underworld) 26

P

Palestine peninsula 31, 35, 44
Paul (apostle) 28, 32-3, 39-40, 48, 75, 83, 118, 124, 134, 152, 171, 173, 186, 198-9, 201-2
Pauline Christianity 186
peninsula *see* Palestine peninsula
Pentateuch 39, 81, 124
Peter (apostle) 28, 83, 128, 141, 144, 172, 176, 178, 202
Pontius Pilate 45, 124, 146, 158, 183
pope 49-50, 207
prophets 19, 87, 96, 110, 113, 117-18, 121, 145-6, 158, 184-6, 189
provinces 35, 45, 113, 164

Q

quest 93, 103, 129, 209
questionable text 183, 189

R

Rabbi   124-5, 159, 162, 166-7
reformation crusade   31, 48-9
religions   17, 27, 29, 33, 45, 48, 50, 86, 122, 127, 168, 178, 197, 206
religious agenda   36, 38, 173
religious ambiguities   37-8, 41, 56, 178-9, 183
resurrection *see under* Jesus Christ
revelations   29, 75, 109
reverence   173
righteousness   22, 96
Roman Empire   5, 35-7, 42-7, 49, 118, 124, 138, 159, 164, 199, 207
ROMANS   43, 51

S

saints   187
Satan *see* devil
sex   63, 65, 125, 177, 179-81, 185
Son of God (*see also* Jesus Christ)   29, 100, 119, 122, 130, 132-3, 149, 165, 176, 179, 191
Son of Man   119-20, 122, 128-9, 131, 141, 143 *see also* Jesus Christ
soul   10, 86, 93, 131, 176
spirits   20, 34, 45
Synoptic:
  Bible   33, 160, 178
  Gospels   165
  scriptures   165, 171
  view   22, 30, 40, 112, 151, 154, 171, 173

T

Ten Commandments   17, 63, 68, 86, 92, 96, 104, 107, 116, 119
territory   43-4, 82, 94-5, 105

testament *see* New Testament; Old Testament
theologians   24, 128, 149, 171
theology   39
Theotokos   34, 37, 67, 113, 171
Tiberius (Roman emperor)   45-6, 118
Timothy   51, 176, 206

U

universe   11, 16, 22, 33, 41, 53-4, 56, 65, 67, 72-4, 82, 91, 150-3, 188-9, 206-8

V

virgin   89, 105, 110-11, 113, 136-7, 152, 157-8, 181, 183-5

W

witches   9, 19, 206
world   20, 27-8, 34, 41, 48, 53-5, 66, 68, 82, 91, 97-9, 155-6, 192-3, 207-8, 211

Y

Yahweh   26-7 *see also* God

Z

Zachariah (John the Baptist's father)   135-7

www.ingramcontent.com/pod-product-compliance
Lightning Source LLC
Chambersburg PA
CBHW021100080526
44587CB00010B/318